Remedial and Analytical Separation Processes

This book describes a comprehensive, integrated view of separation science backed by discussions about simple extraction and partition processes to give a better understanding of advanced techniques like chromatography and membrane separations. It paves the way for an understanding of the fundamental physical and chemical phenomena involved in separations and a concise overview of transport reactions. A chapter dedicated to phytoremediation gives an understanding of the various processes involved in the bioremediation of environmental media.

FEATURES:

- Provides synchronous aspects of the separation process for remediation, including phytoremediation and analysis using chromatography.
- Addresses basic separation techniques for water solutions.
- Discusses mechanistic views of various separation processes.
- Includes the mechanism of separation using membranes and sorbents.
- Helps the reader understand the connection between the different discrete separation processes.

This book is aimed at senior undergraduate and graduate students in environmental engineering and analytical chemistry.

Remedial and Analytical Separation Processes

Jayshree Ramkumar and A.K. Tyagi

CRC Press
Taylor & Francis Group
Boca Raton London New York

CRC Press is an imprint of the
Taylor & Francis Group, an **informa** business

First edition published 2025
by CRC Press
2385 NW Executive Center Drive, Suite 320, Boca Raton FL 33431

and by CRC Press
4 Park Square, Milton Park, Abingdon, Oxon, OX14 4RN

CRC Press is an imprint of Taylor & Francis Group, LLC

ISBN: 978-1-032-45885-4 (hbk)
ISBN: 978-1-032-58093-7 (pbk)
ISBN: 978-1-003-44251-6 (ebk)

DOI: 10.1201/9781003442516

Typeset in Times
by Apex CoVantage, LLC

Contents

Preface

Human beings and the environment, being interconnected, tend to affect each other in more than one way. Constant technological growth is bound to result in environmental degradation. In order to remove these toxic species from environmental media like ground water and to achieve remediation, separation is essential. Separation is a method by which a mixture or solution containing different components is resolved into individual components, leading to the enrichment of a particular phase. Separation can be either analytical or preparative in nature and can be carried out at a lab scale or an industrial level, respectively. In analytical methods, separation is utilized to segregate the required component with a high degree of purity to enable highly accurate and precise determination. The preparative separation aims at achieving a component with a high degree of purity for further applications. The latter can be understood to play a role in remedial applications.

In this book, an attempt is made to give a comprehensive and integrated view of separation processes. The initial chapters are dedicated to discussions on the use of these procedures for remediation purposes. Each chapter discusses the basic concepts of the separation process, followed by their applications for the removal of toxic species. The first chapter elaborates in general about environment and its effect on mankind upon degradation. The second chapter gives an idea of the concept of separation, starting with simple techniques which pave way for the comprehension of advanced techniques. The book is intended to provide an understanding of the fundamental physical and chemical phenomena involved in membrane separation processes. A chapter dedicated to phytoremediation is aimed to provide an understanding of this imminent technique found to be suitable for bioremediation. However, since the technique is taking baby steps, not much work has been reported, and no clear-cut strategy can be obtained. In addition to the different techniques adopted for the removal of toxic species, the book also contains a chapter dedicated to chromatography. This was found to fit within the scope of this book as this analytical methodology is based on separation. The book can be used to understand the basics and also as a ready reference to carry out further research.

Due care has been taken to minimize errors, but some could have occurred due to unintentional oversight. We shall be appreciative of the readers for bringing such inadvertent mistakes to our notice. An attempt has been made to achieve a balance in the contents to make it useful to a large spectrum of readers including students and researchers.

About the authors

Dr. Jayshree Ramkumar obtained her MSc (analytical chemistry) in 1993 from Madras University, Chennai, and joined Bhabha Atomic Research Centre (BARC) Training School, Mumbai, in the same year. After completing a one-year orientation course in nuclear science and technology, she joined the Analytical Chemistry Division of BARC in 1994. She is also an associate professor (Chemistry) at Homi Bhabha National Institute (HBNI), Mumbai. She has been involved in research in the areas of separation science and analytical chemistry. Her PhD was on the studies using Nafion membranes and bulk liquid membrane for achieving separation of various species like metal ions and organic compounds. She was awarded MANA fellowship for carrying out postdoctoral research at the National Institute for Materials Science (NIMS), Tsukuba, Japan. Her work involved the synthesis of mesoporous materials for applications as sorbents for the removal of toxic species. She has been collaborating with universities through the BRNS and AERB projects as the principal collaborator. She is the reviewer of many project proposals of different funding agencies for their suitability for funding. She is a member of the Doctoral Committee of HBNI, editorial board of journals, and serves as an external examiner for PhD students of different universities. She has more than 75 publications, including papers in international journals and book chapters, to her credit.

Dr. A.K. Tyagi joined Chemistry Division, BARC in 1986 through BARC Training School and received PhD degree in 1992. He did postdoctoral research at Max-Planck Institute, Stuttgart, Germany (1995–96). He occupied various positions at BARC such as Director, Chemistry Group, Director, Bio-Science Group, Head, Chemistry Division, etc. He was a Distinguished Scientist, Department of Atomic Energy (DAE). Presently, he is Dean and Senior Professor, Homi Bhabha National Institute (HBNI), Mumbai and Honorary Professor at JNCASR, Bengaluru. His research interests are in the field of anomaterials, functional materials, nuclear materials, metastable materials, and hybrid materials. He has to credit his 47 PhD students and about 650 papers in journals. He has been conferred with a number of awards such as DAE-Homi Bhabha Science and Technology Award; DAE-SRC Outstanding Researcher Award; MRSI Medal; MRSI-ICSC Materials Science Senior Award; MRSI-CNR Rao Prize in Advanced Materials; MRSI Distinguished Materials Scientist of the year award; CRSI Bronze Medal; CRSI-CNR Rao National Prize in Chemical Sciences; CRSI-Silver Medal; RajibGoyal Prize in Chemical Sciences; Metallurgist of the Year award from Ministry of Steel, GoI; National Prize in Solid State and Materials Chemistry from JNCASR; ISCA Acharya PC Ray Memorial Award; NETZSCH – ITAS Award by Indian Thermal Analysis Society; DN Agarwal Memorial Award from Indian Ceramics Society; Gold medal of Chirantan RasayanSanstha. He is Fellow of several national and international science academies such as National Academy of Sciences, India (FNASc); Indian Academy of Sciences (FASc); Indian National Academy of Engineering (FNAE); Indian National Science Academy (FNA); Royal Society of Chemistry (FRSC); and The World Academy of Sciences (FTWAS). Recently, he was conferred with the restigious Vigyan Shri Award by Hon'ble President of India.

1 Introduction to environment alteration and reshaping

1.1 WHAT IS ENVIRONMENT?

Man and environment are connected to one another and cause changes in each other. Environment gives life to mankind, but the changes caused by man lead to degradation of environment. The relationship between man and the environment has been established in the early periods itself wherein humans lived in sync with the nature and followed the laws of nature. The knowledge of environment through systematic education can help work towards a sustainable future.

The term environment emanated from *environia* (French word), which means "to surround". Therefore, it is symptomatic of both abiotic (physical/non-living) and biotic (living) surroundings within which humans live [1]. Thus, both environment and humans are intertwined in a complex manner, and the environment controls every aspect of life. It is very obvious that the interaction between humans and environment is more prominent as compared to that of other species. Accordingly, environment can be defined as the surrounding or conditions in which humans and other living species are present and can refer to the elements of the physical and biological world and the interactions between them. Consequently, environment can be considered as the conditions and interactions in toto at a given point of time and space. As the environment consists of atmosphere, hydrosphere, lithosphere and biosphere with the main components (soil, water and air), it is clear that there will be a direct influence (natural, artificial, social, biological and psychological) on the lives of human beings. Environment can be categorized into micro and macro environments and also as non-biotic and biotic. Macro environment is the local surrounding of an organism and involves physical and biotic conditions. The physical environment includes abiotic factors like temperature, light, rain and soil, while biotic indicates all living forms. Environment is productive (provides raw materials for various industries), aesthetic and provides an understanding of the value and importance of the surroundings. Therefore, the need to preserve and reuse the resources becomes a major facet of survival.

1.2 CHANGING DYNAMICS OF ENVIRONMENT

The incessant expansion of scientific know-how leads to immense apprehension on the upshot of the milieu, thus necessitating the formulation of the modus operandi of the removal of the different obnoxious waste generated during these processes. Environment is all of those that are contiguous to a living organism. Due to the

DOI: 10.1201/9781003442516-1

unremitting development, environment is never static and results in pollution due to the introduction of unwanted species, normally referred to as either a contaminant or a pollutant. But there is a subtle difference between the two terms. A contaminant can be defined as a species which should not be present at a particular place. When a contaminant is present at levels that cause damage to human life, it is defined as a pollutant [2]. The unwanted species can be natural or man-made or can also be biological or chemical in nature [3]. Therefore, it can be easily understood that *not every contaminant is a pollutant, but every pollutant is definitely a contaminant* [4]. The famous quote of Paracelsus, the father of toxicology, *sola dosisfacitvenenum* is the adage intended to indicate a basic principle of toxicology: "The dose makes the poison" [5]. Thus, it is clear the concentration decides the toxicity of a substance. From the famous quote of B. F. Skinner, "The environment will continue to deteriorate until pollution practices are abandoned", it is quite clear that the present practices of mankind need to be altered to reduce the pollution in the environment.

1.3 WATER: THE FLOW OF LIFE

Water is one of the important components of the environment which is of great importance to mankind. According to the famous quote of Vriddha Chanakya, water is one of the gems of Mother Earth. Acharya Vagbhata also quoted that water is the life of all living creatures on this earth, and the major portion of earth contains water. It was understood everyone needs water irrespective of their health status [6]. Water, known as Ap in Sanskrit, is the fourth among the five elements of the universe (panchamahabhuta). The various forms of water like snowfall and dew have also been mentioned. Since water is very important for the sustenance of life, even a slightest deviation in its characteristics becomes quite a challenge for mankind [7]. The increasing stress on water environment leads to decreased availability of clean water and also a greater damage to aquatic life. Preservation of our water environment must be integrated with development to make it sustainable. This therefore needs the development of effective wastewater treatment protocols to be integrated with policies for continuous development.

The total volume of water on earth is estimated at 1.386 billion km³ (333 million cubic miles) and is present in various forms as shown schematically in Figure 1.1, based on the literature reported [8]. Of the total amount of water, 97% is saline (salt water) and only 3% is fresh water, and out of this, only 0.3% is present on the surface and is accessible for use.

The water on the planet is in a state of continuous motion and is temporarily stored as different water bodies or reservoirs (rivers, lakes, ponds and ice caps) on the earth's surface and as groundwater below the surface [9]. Residence time (time of water present in these reservoirs) is not constant and is affected by various conditions. This can be understood easily from the water cycle, also referred to as the hydrological or hydrologic cycle [9]. Hydrological cycle denotes the sequential environmental processes which lead to the continuous circulation of water on the earth's atmospheric system [9]. Water (in solid or liquid form) is converted to vapour, which traverses substantial distance from its source prior to recondensation. Although the total amount of water within the cycle remains essentially constant, its distribution

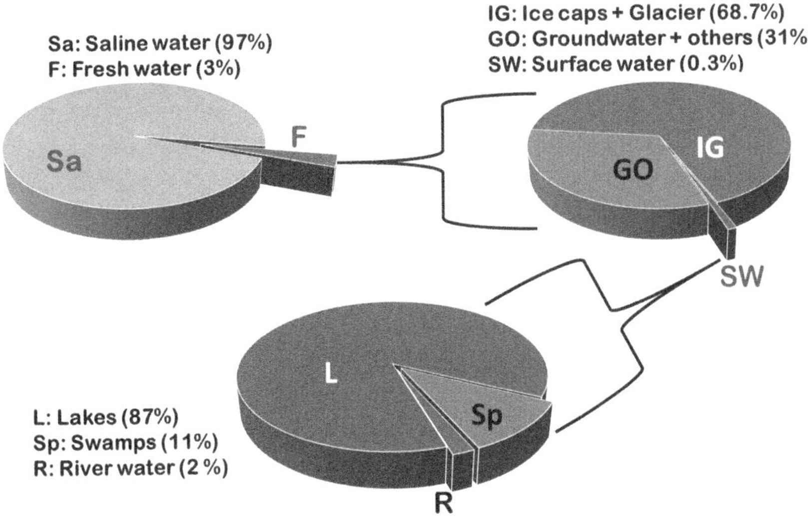

FIGURE 1.1 Schematic representation of the distribution of different types of water on earth

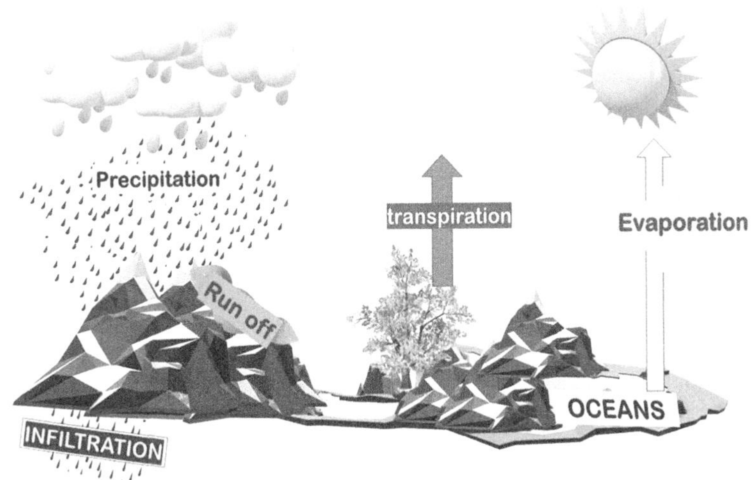

FIGURE 1.2 Schematic representation of the hydrological cycle

among the various processes changes continuously. Figure 1.2 gives a schematic representation of the hydrological cycle. Generally, the starting point of the water cycle is considered to be the starting point of the hydrological cycle (due to large quantities of water). It is understood that the hydrological cycle is a complex process involving different aspects which are linked to each other. It is now understood that the earth's freshwater supply is solely due to the hydrological cycle involving different stages, namely, evaporation, condensation, runoff, streamflow, infiltration and transpiration (evaporation from leaves). Water from the ocean evaporates, condenses into minute particles suspended in the air, and finally precipitates on the earth's surface due to condensation of the vapour. The vapour condenses due to the movement of warm air into a colder zone horizontally or vertically above the earth's surface or due to cooling when it moves across the mountains. A part of this vapour can directly form ice and snow by the process of sublimation. Thus, the processes of condensation and evaporation are together referred to as precipitation. The falling of rain on terrestrial areas and capture by vegetation is known as interception, and the possibility of this water being evaporated again is quite high (due to the exposure of a large area to wind). The remaining amount which falls on the earth can either be trapped within the soil surface (infiltration) or can flow as streams (surface runoff). Runoff is the total of amount of water flowing as stream due to surface runoff and the amount of water present as ground water. Infiltration is the percolation of water into the earth's crust by different modalities. The recharge of underground water occurs due to run-off of water. The hydrological cycle may have a long cycle or many short cycles. In the short cycles, water evaporates from the marine or freshwater system, condenses immediately and precipitates in the same area, and this cycle can keep occurring. In the long cycle, water evaporates from the oceans, forms clouds that move inland and then precipitates. This water will return to the ocean. Thus, the precipitation can occur quite close to or far away from the original source. It is estimated that the water remains in the atmosphere for about 10 days. The hydrological cycle is essential to maintain the balance in the ecosystem and provide fresh water. The sole source of land water is atmospheric precipitation. It is estimated some of these flows into the oceans while the majority is evaporated.

India has an area of $3,287,590$ km^2 with a 7,500-km-long coastline [10]. Throughout the country, there are huge variations in geographical features and climate. The river network is quite well spread across the entire country. The major rivers of India are divided based on the ocean into which they flow [11]. The rivers Brahmaputra, Yamuna, Ganga (with its main tributaries Ramganga, Gangan, Kali or Sharda, Gomti, Yamuna, Chambal, Betwa, Ken, Tons, Ghaghara, Gandaki, BurhiGandak, Koshi, Mahananda, Tamsa, Son, and Bagmati), Meghna, Mahanadi, Godavari, Krishna (and its main tributaries) and Kaveri flow into the Bay of Bengal, while the rivers Narmada, Tapi, Sindhu, Sabarmati, and Purna flow into the Arabian Sea. The river basins are categorized based on the catchment area covered. There are 13 major river basins with a catchment area of more than $20,000$ km^2, contributing to 85% of the total surface flow [11]. The major river basins are Brahmaputra, Ganga (including Yamuna sub-basin), Indus (including Satluj and Beas sub-basins), Godavari, Krishna, Mahanadi, Narmada, Cauvery, Brahmini (including Baitarni sub-basin), Tapi, Mahi, Pennar, and Sabarmati. There are 48 medium river basins with a catchment area in the range of $2,000$–$20,000$ km^2, which contributes about 8% to the total surface flow. There are 52 minor river basins with a catchment area of less than $2,000$ km^2, contributing to about 9.6% of the total surface flow. There are few desert rivers, which flow for some distance and are lost in deserts, while some areas are completely arid (evaporation equals rainfall) with no surface flow. The medium and minor river basins are mainly in the coastal areas. The rivers on the east coast is about 100 km due to the large distance between the land and the sea. The rivers in the west coast are much shorter as the width of the land between the sea and mountains is about 10–40 km.

Despite nature's bounty, water paucity results in deterioration of water quality in aquatic resources and hence is an issue of national concern. The exponential increase in population along with rapid technological development has affected both the quantity and quality of water in India. The main aspect of urbanization is the skewed distribution of more than 25% of the population within the metro cities alone. The unregulated growth of urban areas has led to increased pollution, and the situation warrants immediate redressal through radically improved water resource and water quality management strategies. The availability of freshwater resources is declining in India [12] as the requirements are increasing many folds.

1.4 POLLUTION: MAKING ENVIRONMENTAL DUSTBINS

The introduction of unwanted matter from various sources into large water bodies leads to the degradation of the quality of water and, therefore, has a great impact on life [13]. Jacques Cousteau, a French naval officer, scientist, and conservationist, said: "We forget that the water cycle and the life cycle are one". This is a very true statement because water is the flow of life, and all life's sustenance depends on water quality. However, ongoing rapid industrial and commercial developments have deteriorated water quality, so poor as can be understood from the quote of Jacques Cousteau: "Water and air, the two essential fluids on which all life depends, have become global garbage cans". The contaminants found in groundwater can be physical, inorganic or organic chemicals, bacteriological and radioactive in nature. The survival of mankind depends on the three basic natural resources water, air and soil. The concept of pollution (cause and effects) has been discussed in 5,000-year-old Indian classics like Sushruta Samhita and Charaka Samhita, wherein many methods of purification have also been discussed [6, 14]. According to Susrutha (the Father of Indian medicine), there are six types of pollution (dosha; sparsa, rupa, rasa, gandha, virya, and vipaka) which have negative effects on man. It has observed that there is a continuous increase in the demand for water due to the continuous increase in population, urbanization and industrialization expansion combined with the increased need for agricultural produce. The per capita availability of water (estimated by dividing the annual average water availability by population) is also continuously decreasing. Therefore, sustenance and management available water become a huge challenge. Furthermore, statistics related to the adverse effects of consumption of contaminated water is alarming, adding further complexity to the water stress. Although Asia has 36% of the available fresh water on the earth, the advantage is offset by its large population (over 60% of the world population), leading to water scarcity. Australia and the oceans have plenty of water followed by North, Central, and South Americas. Africa is still in a better position than Asia [15]. It is projected that by the year 2050, a huge percentage of people would be affected. The demand for water in India is projected to overtake the amount available. The main increase will be from the domestic sector. The National Commission for Integrated Water Resources Development (NCIWRD) has indicated that the water demand in the domestic and industrial sectors would increase substantially. India is not a water deficit country, but neglect and lack of monitoring have resulted in water stress. The concern regarding environmental degradation at the global level was first emphasized at the UN Conference in Stockholm in June 1972 [16]. Thereafter, topics like environment, sustainability and water quality became the key aspects. Various studies suggested that not only the need for water was more in urban areas than in rural areas, the quality of water discharged in urban areas was also very poor.

Water quality is defined as a measure of the suitability of water for a particular use based on select physical, chemical, and biological characteristics [17,18]. It is also a measure of the conditions relative to the requirements [19,20]. Water can be classified into different categories, namely, potable water, palatable water, contaminated (polluted) water, and infected water, based on its quality [21]. Potable water is water which is safe to drink, pleasant to taste and usable for domestic purposes [21]. Palatable water is aesthetic but contains some contaminants at very low concentrations such that they do not cause any adverse reactions to human life. Contaminated water contains unwanted physical, chemical, biological, or radiological substances, and it is unfit for drinking or domestic use, while polluted water causes adverse health effects [22]. Infected water is polluted with pathogenic organisms. Physical, chemical, and biological parameters are used to check and ensure the quality of water [23,24].

(i) ***Physical parameters of water:*** The different physical parameters that are generally checked are turbidity, temperature, colour, taste and odour, suspended solids, and electrical conductivity.

Turbidity is the cloudiness of water [25] caused by suspended particles (clay, silt, organic material, and plankton), making it unfit for consumption. It leads to increased pretreatment cost [26], can hinder the disinfection process [27] (for removing toxic microorganisms), can prove to be harmful to aquatic and human life and tends to increase the concentrations of toxic heavy metal ions or organic compounds [28]. The presence of turbidity increases temperature and thus reduces dissolved oxygen (DO). Turbidity is measured using nephelometric turbidimeter and is expressed in units of NTU or TU (1 TU = 1 mg/L of silica in suspension) [25]. Turbidity of more than 5 NTU is visible, and the value is greater than 100 NTU for muddy water [25]. Groundwater normally has very low turbidity because of the natural filtration that occurs as water penetrates through soil [29].

Temperature affects the different characteristics of water, including oxygen content and metal ion biosorption [30,31]. Water at a temperature of 10–15°C is found to be most palatable for many people [32]. An increase in the temperature of seas and rivers will cause great harm to aquatic life.

Water colour can be caused by the decay of organic matter and the presence of certain inorganic species. The presence of colour is not acceptable not only due to aesthetic reasons but also because it can prove to be toxic to human health [33]. Water colour is measured by comparing the water sample with standard odour solutions or odoured glass disks [25]. One colour unit is equivalent to the colour produced by a 1-mg/L solution of platinum [potassium chloroplatinate (K_2PtCl_6)] [25]. Colour is graded on scale of 0 (clear) to 70 odour units. Pure water is colourless, which is equivalent to 0 colour units [25].

Taste and odour of water can be altered due to the presence of organic materials, inorganic compounds, or dissolved gasses which may be introduced from natural, domestic, or agricultural sources [34]. The numerical value of odour or taste is determined quantitatively by diluting a known volume of a sample with a measured volume of distilled water (B) till the total volume is 200 mL [35]. Threshold odour number (TON) and threshold taste number (TTN) are units of odour and taste, respectively. These are whole numbers indicating how many dilutions it takes to produce odour-free water.

When water is filtered, the suspended solids are retained on the filter, while dissolved solids tend to pass through the filter along with water [25]. When this sample is evaporated, the dissolved solids are obtained as residue. The total concentration of dissolved solids, referred to as total dissolved solids (TDS), is used to classify water. The TDS values in fresh, brackish and saline waters are <1500 mg/L, 1500–5000 mg/L, and >5000 mg/L, respectively. The residue of TSS and TDS after heating to dryness for a defined period of time and at a specific temperature are defined as fixed solids. Volatile solids are solids lost on ignition (heating to 550°C) [25].

Electrical conductivity (EC) of water is a measure of the ability of water to conduct current [35] due to the presence of charged species and is expressed as milliSiemens per metre (mS/m). The values of conductivity for ultra-pure water, drinking water and sea water are 5.5×10^{-6}, $0.005 - 0.05$, and 5 S/m, respectively. This value can be correlated with the TDS value.

(ii) *Chemical parameters of water quality*

One of the important chemical parameters which decides the water quality is its pH. Pure water is nearly neutral (pH ~7.0 at 25°C), while rainwater is slightly acidic (pH ~5.6) due to the dissolution of carbon dioxide [36]. The pH value specified for drinking water is $6.5 - 8.5$ [36]. A very high pH value renders a bitter taste to water and also makes the disinfection process ineffective. A high pH value makes water corrosive and causes leaching out of different species. The changes in pH affect the aquatic life (e.g. decreased hatching, and irritation and damage of membranes). Similarly, alkalinity can also be used as a pollution indicator. The concentration of certain inorganic ions can be used as a pollution indicator. Nitrogen in solution can be present as organic nitrogen compound, ammonia, nitrite, and nitrate [25]. Hardness of water indicates the presence of high concentrations of minerals (bicarbonate and sulphate salts of calcium and magnesium), which can cause deposits on the piping system. Temporary hardness is caused due to carbonates and bicarbonates, while permanent hardness is due to the presence of sulphates and chlorides. Temporary hardness unlike permanent hardness can be removed by boiling of water. The unit of hardness is mg/L of $CaCO_3$, and the values for soft, moderate, hard and very hard water are <50, 50–150, 150–300, and >300 mg/L, respectively. A hardness value of up to 500 mg/L is considered safe, and a value above this shows laxative effects [37, 38].

DO is an important parameter indicating water quality [25]. Water with higher DO is considered to be of very good quality. If the DO concentration decreases to less than 4 mg/L, all fish will die, while a DO concentration less than 2 mg/L leads to death of all organisms, and the water is called septic water. The ideal DO concentration value is between 8 and 10 ppm, and the optimum value is 9 ppm. The DO concentration is very sensitive to water temperature [35]. Cold water has more DO than warm water (14.6 ppm at 0°C; 9.1 ppm at 20 C) as increased temperature leads to denaturation of enzymes, thus leading to a decreased photosynthesis rate [39].

(iii)   Biological parameters are the aquatic life in water which serves as an indicator of water quality and can be estimated by calculating the species diversity index (SDI) [40]. The multiplication rates of different species are different and also sensitive to the external conditions. The quality of water and its level of pollution are marked by various indicators, including pH, BOD, COD, and total coliform counts. Biochemical oxygen demand (BOD) refers to the decrease in DO due to the activity of bacteria and microorganisms [41] and can be used as a marker. Sewage water will have high BOD values as it contains more organic matter. Food industry waste water contains high organic loading compared to municipal waste water. Therefore, BOD is used as a measure of the power of sewage (high BOD values indicate strong sewage) [42]. It is known that microorganisms take 20 days to completely decompose the organic matter in water [42]. The measure of oxygen required to completely decompose organic matter in a specified volume of water is called the ultimate BOD or BODL. Chemical oxygen demand (COD) is the total measurement of all chemicals in water that can be oxidized, while BOD is the measure of food (organic carbon) that bacteria can oxidize. COD indicates the concentration of biodegradable and non-biodegradable substances [41] and is used as a measure of organic pollutants in water and is expressed in milligrams per litre (mg/L) or parts per million (ppm). It is determined using a combination of strong oxidizing agents (potassium dichromate), sulphuric acid and heat [25]. The most common application of COD is in quantifying the amount of oxidizable pollutants found in water. The average values of COD and BOD are in the ranges of 300–500 mg/L and 200–300 mg/L, respectively, while both these values are in the range of 1,000 to >1,00,000 mg/L in food industry effluents [42]. COD values are always higher than BOD values for the same sample [42]. BOD measures the amount of oxygen required by aerobic organisms to decompose organic matter, and COD measures the amount of oxygen required to decompose organic and inorganic constituents present in waste water by chemical reaction. Hence, the value of COD is greater than that of BOD. The ratio of BOD:COD is a good indicator of pollution. It has been observed that the BOD value has a direct effect on the value of DO in rivers and streams. The greater the BOD, the more rapidly oxygen is depleted in the stream. This means less oxygen is available to higher forms of aquatic life. Moderately polluted rivers may have a BOD value in the range of 2–8 mg/L. Rivers may be considered severely polluted when BOD values exceed 8 mg/L [43]. Municipal sewage that is efficiently treated by a three-stage process would have a BOD value of about 20 mg/L or less. The total coliform count gives a general indication of the sanitary condition of a water supply [44]. A positive coliform test means possible contamination and a risk of waterborne disease. A positive test for total coliforms always requires more tests for faecal coliforms or *E. coli*. The EPA maximum contaminant level (MCL) for coliform bacteria in drinking water is zero (or no) total coliform per 100 ml of water.

When the parameters of water are altered due to various reasons, it is said to be polluted and considered not potable. The water quality requirement for different uses of water indicates the purity of water. It is crucial that drinking water should be very pure. In order to set the standard of water quality, knowledge of its use becomes important. According to the Central Pollution Control Board (CPCB) of India, there are five classes of water, as discussed below. Class A is drinking water source which can be used without conventional treatment and with only disinfection. The pH of the water is in the range of 6.5–8.5, with DO and BOD values of 6 and 2 mg/L, respectively. The total coliform organism should be less than 50 MPN/100 mL. Class B water is used for outdoor activities and should have pH in the range of 6.5–8.5. The DO is greater than 6 mg/L, while the BOD is around 2 mg/L and the total coliform organism is less than 50 MPN/100 mL. Class C water is a drinking water source pretreated with conventional treatment followed by disinfection. The total coliform organisms should be less than 5,000 MPN/100 mL, and the pH is in the range of 6–9. The DO and BOD values should be around 4 mg/L. Class D water is used for the propagation of wild life, fisheries, etc., and has pH in the range of 6.5–8.5 and DO of more than 4 mg/L with free ammonia of up to a maximum of 1.2 mg/L. Class E water, with pH of 6–8.5, is used for irrigation, industrial cooling, and controlled waste disposal. The last class is that below Class E, wherein water does not meet any of the specifications. For easy understanding and representation on maps, an universally accepted colour code is used. Blue water can be directly used for drinking and industrial purposes. Green water depicts the water present in soil and plants. White water refers to fast shallow stretches of water in a river. White water is formed when the gradient of river being high causes turbulence making the water appear white. Brown or grey water refers to various grades of waste water.

1.5 ORIGINS OF WATER POLLUTION

Water is essential for sustaining life, as the British poet W. H. Auden once noted, "Thousands have lived without love, not one without water". Nevertheless, it is not treated with care. The dumping of waste water without any pretreatment into the environment results in the pollution of all water bodies like rivers, lakes, and oceans. Water pollution is the second most prevalent problem after air pollution. Water is very vulnerable to pollution. Being referred to as "universal solvent", water has the ability to dissolve a large number of species of which many could make it toxic.

Water pollution can be categorized in different ways, namely, based on the type of water polluted or the nature of the pollutant. In the first category, the different types of water bodies can be understood. Groundwater is created when rain falls and seeps into the earth, filling up the cracks, crevices, and porous structures, and is an important natural resource. Groundwater gets polluted with contaminants (pesticides, fertilizers, and waste leachates). Surface water is the water present in oceans, lakes, and rivers and comes from freshwater sources. The major pollutants are nitrates and phosphates from fertilizers and agricultural runoffs. Municipal and industrial waste discharges also result in the toxicity of water. Marine pollution is known to originate from the land (coast or inland), and different chemicals and nutrients are carried from farms and factories by streams and rivers into bays and estuaries and finally into the sea. Marine debris (plastic) is blown by wind or washed via storms, drains, and sewers. Oil spill is another contributing factor to the pollution of oceans. The greatest source of pollution of fresh water is from agricultural run-offs. Sewage water is also one of the major contributors to water pollution.

The different types of pollutants can be classified as biological (bacteria, viruses, worms, and protozoa—added by excreta of animals), chemical (inorganic such as phosphates, nitrates, fluorides, and chlorides, and organic such as phenols, plastics, dyes, pesticides, and chloro compounds), heavy metal ions (cadmium, mercury, copper, zinc, and their organometallic compounds), physical (waste heat from industrial plants causing physical pollution involving changes in the physical properties of water, e.g. turbidity and temperature). The different sources of pollutants are domestic effluents, industrial effluents, surface runoffs, and waste heat. Domestic effluents are those that are discharged into the common public sewerage system and may contain excreta, food residue, detergents, and bacteria. The pollutants may be present as suspended solids (sand, silt, and clay), colloidal particles (large-sized suspended inorganic or organic compounds like faecal matter, bacteria, cloth, paper, and fibres), or dissolved solids (nitrates, ammonia, phosphates, sodium, calcium, toxic metallic ions, and organic compounds). Industrial effluents are discharges coming from industries and contain inorganic pollutants (mercury, lead, copper, arsenic, cadmium, acids, alkalies, and bleaching liquors) and organic pollutants (phenol, naphtha, proteins, cellulose fibres, aromatic compounds, putrescible organic matter) and can be carcinogenic in nature. The main sources of mercury are combustion of impure coal, thermal power plants, chloralkali industries producing caustic soda, thermometers, blood pressure instruments, smelting of metallic ore, and paper and paint industries, while lead comes from smelters, battery industry, chemical and pesticide industries, automobile exhausts, etc. Cadmium is released from electroplating, pesticide, and phosphate industries. The surface runoff from the field introduces pollutants like inorganic fertilizers, pesticides, insecticides, and manures, which are non-degradable, into natural water bodies.

The different sources of water pollution can be categorized as point and non-point sources [45], which are further classified into natural and anthropogenic sources. Point source pollution is when pollutants (domestic/industrial waste waters) are directly discharged into freshwater bodies from well-defined sources. The discharge can be easily monitored and controlled. Non-point sources are spread over a large area, leading to indirect introduction of pollutants (runoffs from agricultural farms, construction sites, abandoned mines, and solid waste disposal sites) and making it difficult to control. Natural source refers to the sudden increased concentration of natural substances, e.g. siltation, which occurs due to haphazard deforestation (the solid is loosened and displaced by flowing water). Anthropogenic sources are the result of human activities from both increased population and technological growth leading to increased domestic and industrial waste.

1.6 SCENARIO OF WATER POLLUTION: INTERNATIONAL AND NATIONAL

The problem of water pollution is a global concern, with countries struggling to get clean water and combat fatal waterborne diseases. It is also seen that individuals with decreased immunity are more susceptible to fatality even with comparatively less threatening waterborne diseases. The major water bodies present all over the world [46–68] are affected in different ways, resulting in various hazardous effects on both aquatic life and mankind. The presence of electroplating industries on riverbanks results in high concentration levels of chromium, which can enter the food chain and cause great damage. The discharge from sewage, laundry, agriculture, and municipal activities can make the water alkaline and change the DO values.

1.7 EFFECT OF WATER POLLUTION: HUMAN HEALTH AND CLIMATE

Drinking water standards are the most important parameters when it comes to deciding whether the water is safe or not. This is because the main objective is to select the appropriate methodology to ensure complete removal of water pollutants. The primary aim of water treatment protocols is to produce biologically and chemically safe water which is also aesthetically acceptable and also does not produce corrosion or scaling problems. The standards set for drinking water are regulations set by various regulatory bodies to control the level of contaminants in drinking water. Inorganic substances at very low concentrations are in general important for sustenance of human life, whereas at higher concentrations these tend to prove detrimental. For each and every species, there are specific values given by different regulatory bodies like WHO [69], EU [70], US [71], Canada [72], and BIS, India [73], to prevent different adverse effects to mankind. Aluminium does not show any detrimental effects, but the EU gives a value of 0.2 mg/L. Antimony in water leads to increase in cholesterol, and the values given by different bodies is about 0.006 mg/L. Arsenic being carcinogenic is specified at a value of 0.01 mg/L by various bodies. The presence of chromium causes allergic dermatitis, and the value is about 0.05 mg/L. Lead is a toxic metal ion which leads to delay in the development of foetus and infants and also kidney damage in adults. The limit in water has been specified by various bodies to be about 0.01 mg/L. Mercury is another toxic heavy metal ion which leads to severe kidney damage, and the limit given in water is about 0.001 mg/L. Fluoride is one of the most abundant anions present in groundwater worldwide and creates a major problem in safe drinking water supply [74]. It is always present as inorganic or organic fluorides [75]. High fluoride concentrations can build up in groundwater present in aquifers, while surface and shallow groundwater usually have low concentrations. The fluoride concentration is dependent on climatic condition [76, 77]. Excess fluoride in water causes bone diseases and mottled teeth, and values given as specifications by different bodies range from 1 to 4 mg/L.

Dyes have been used extensively since ancient times. Dyes exist in nature, but with time, researchers have gained expertise in synthesizing dyes [78]. According to Witt theory [79], dyes have a chromophore group which imparts odour to the dye and auxochromes that intensify the odour. The most important auxochromes are amine ($-NH$), carboxyl ($-COOH$), sulphonate ($-SO_3H$), and hydroxyl ($-OH$). Depending on the application method, most of the dyes can be classified into acid, basic, direct, reactive, disperse, vat, mordant, and sulphur [80].

Textile industries devour huge amounts of water [81] and chemicals with miscellaneous chemical composition [82] and cause great environmental impact [83]. There are very large numbers of dyes available, with over 10^5 tonne of dye-stuff production annually. The main obstacle with dyes are that they are not easily degradable. The Ecological and Toxicological Association of the Dyes and Organic Pigments Manufacturers (ETAD, established in 1974) intends to minimize environmental damage, protect users and consumers, and look into the concern regarding toxicity of dyes. The highest rates of toxicity were found amongst basic and direct diazo dyes [84]. Water-soluble reactive and acid dyes cause great damage [85] as removal is difficult using conventional methods. Non-ionic dyes are disperse dyes which do not ionize in solution and are carcinogenic in nature. The presence of fused ring structures reduces the degradation of anthraquinone dyes [86]. The resistance of these dyes to degradation leads to their accumulation in the environment or partial degradation, producing dangerous by-products. The xenobiotic nature of the dyes affects the structure and functions of ecosystems. Continuous exposure to metal-complexed dyes usually used in textile industries causes severe damage to aquatic biota and to human health [87]. Textile dyes can cause dermatitis, disorders of the central nervous system, or enzymatic deactivation [88]. Oral ingestion or inhalation leads to skin and eye irritations, and textile industry workers were found to suffer from contact dermatitis, allergic conjunctivitis, rhinitis, occupational asthma, and other allergic reactions [88]. This is due to the formation of a conjugate between human serum albumin and the reactive dye producing immunoglobulin E (IgE) antibodies, which combine with histamine [89]. It is established that the dyes can cause mutation [89]. Azure B, a widely used dye, is easily partitioned to the lipid membrane of the cells [90] and tends to intercalate with the helical structure of the DNA [91] and duplex RNA [92]. This dye shows a great degree of cytotoxicity by acting as a reversible inhibitor of an intracellular enzyme of the central nervous system, namely, monoamine oxidase A (MAO-A) [93]. The enzyme inhibition tends to affect human behaviour and cellular redox homeostasis [94, 95]. It is seen that Disperse Red 1 and Disperse Orange 1 dyes cause mutations due to the formation of DNA adducts and are carcinogenic [96]. Sudan I dye (Solvent Yellow 14) is an azo-lipophilic compound used in different industries and illegally in foods like paprika [97, 98]. As the dye is assimilated into the body, enzymatic transformation into carcinogenic aromatic amines takes place through the action of the intestinal flora [99]. Basic Red 9 dye (used in different industries like textile, leather, paper, and ink) is carcinogenic and toxic. It is transformed under anaerobic conditions to carcinogenic aromatic amines, resulting in allergic dermatitis, skin irritation, mutations, and cancer [100]. Crystal violet dye (a cationic triphenylmethane group dye) has a very intense odour and causes mitotic poisoning, chromosomal damage, and cancer [101].

The increase in population and constant development pollute the water bodies, posing threats to humans and the aquatic ecosystem. The effect of water pollution on climatic changes and the subsequent effect on the hydrological cycle are being studied

[102]. The two terms weather and climate are often used in lieu of one another, but there is a difference between the two terms. Weather is defined as the current atmospheric condition (temperature, rainfall, wind, or humidity) in a given place. Climate is the average weather of a given place or a region. It can be understood as the statistical information of weather variation of a place for a long period of 30 years [103]. Climate variability refers to variations in the mean state and other climate statistics on all temporal and spatial scales beyond those of individual weather events. Variability may result from natural internal processes or from anthropogenic processes [104]. Various scientific communities like the National Aeronautics and Space Administration (NASA), the National Oceanic and Atmospheric Administration (NOAA), and the Environmental Protection Agency (EPA) of the United States have studied this in detail. It is also reported that freshwater resources can be severely affected as evident from the excerpt from the executive summary of a paper, which states, "Observational records and climate projections provide abundant evidence that freshwater resources are vulnerable and have the potential to be strongly impacted by climate change, with wide-ranging consequences for human societies and ecosystems" [105]. Thus, it is quite evident that the situation is indeed challenging as a vicious cycle is created between water pollution and climate change. This is because a sequence of reciprocal cause and effect arises resulting in aggravation of the situation, leading to inexorable deterioration of the situation due to the effect on the water and hydrological cycles. Climate change can be tracked by various indicators, which can be physical, ecological, or societal in nature. The indicators are the values or trend of the values over a given area for a specified time period. The use of indicators can help in understanding the impacts of climate change on the environment and mankind [106].

1.8 ENVIRONMENTAL SUSTAINABILITY: CHALLENGING BUT NOT IMPOSSIBLE

Sustainable development implies the availability of good living conditions by providing solutions to various challenges without any threat to both humans and environment. It involves the protection of resources from further damage by integrating researchers and experts from various fields to understand the challenges and also make correct policy decisions. Environmental sustainability is all encompassing as it is both interdisciplinary and multidisciplinary in nature. Here it is appropriate to quote Sidney Sheldon, who says, "Try to leave the Earth a better place than when you arrived". Therefore, wastewater treatment is a very effective mode of achieving such sustained good living conditions. There are different methods available for water treatment, but different parameters like ease of operation and cost-effectiveness dominate the choice of treatment method.

1.9 CONCLUSIONS

From the earlier discussions, it is seen that in order to comprehend the effect of toxic species on human health, different important aspects need to be integrated. The first step involves the assessment of the nature and amount of toxic species present. This is an important branch of science known as analytical chemistry. Experts in the medical field are required to give an understanding of the hazardous effects of the various species present in water bodies in order to place a maximum limit to the concentration of these species. Therefore, an integrated approach is needed to identify the problem and find ways to solve it. This form of integrative approach is known as interdisciplinary approach. Interdisciplinary research (IDR) is a mode of research by teams or individuals that integrates information, data, techniques, tools, perspectives, concepts, and/or theories from two or more disciplines or bodies of specialized knowledge to advance fundamental understanding or to solve problems whose solutions are beyond the scope of a single discipline or field of research practice.

REFERENCES

1. D.L. Johnson, S.H. Ambrose, T.J. Bassett, M.L. Bowen, D.E. Crummey, J.S. Isaacson, D.N. Johnson, P. Lamb, M. Saul, and A.E. Winter-Nelson, *J. Environ. Qual.* 26 (1997) 581–589.
2. T.M. Pankratz, *Environmental Engineering Dictionary and Directory*, CRC Press, 2000.
3. https://www.epa.gov/ccl/definition-contaminant.
4. P.M. Chapman, *Environ. Int.* 33 (2007) 492–501.
5. J.F. Borzelleca, *Toxicol. Sci.* 53 (2000) 2–4, https://doi.org/10.1093/toxsci/53.1.2.
6. H. Kumari, *Int. J. Res. Anal. Rev. (IJRAR)* 5 (2018) 173y–179y.
7. W.J. Cosgrove and D.P. Loucks, *Water Resour. Res.* 51 (2015) 4823–4839, https://doi.org/10.1002/2014WR016869.
8. I.A. Shilomanov, *"World Fresh Water Resources": Chapter 2 in "Water in Crisis"*, Ed. P. Gleick, Oxford University Press, 1993.
9. V.J. Inglezakis, S.G. Poulopoulos, E. Arkhangelsky, A.A. Zorpas, and A.N. Menegaki, Chapter 3 – Aquatic Environment, In *Environment and Development*, Ed. S.G. Poulopoulos and V.J. Inglezakis, pp. 137–212, Elsevier, 2016.
10. S.K. Kurunthachalam, *Hydrol. Current Res.* S10 (2013) 001, https://doi.org/10.4172/2157-7587.S10-001.
11. V. Jain, N. Karnatak, A. Raj, S. Shekhar, P. Bajracharya, and S. Jain, *Wat. Sec.* 16 (2022) 100118.
12. C.P. Kumar, *J. Sci. Eng. Res.* 5 (2018) 137–147.
13. N. Khatri and S. Tyagi, *Front. Life Sci.* 8 (2015) 23–39.
14. L.C. Senan, S. Balaji, and R. Tripathy, *J. Ayurv. Integ. Med. Sci.* 4 (2019) 75–78.
15. T. Ahmad and T. Bhat, *J. Environ. Earth Sci.* 4 (2014) 67–72.
16. P. Boudes, United Nations Conference on the Human Environment, In *Green Ethics and Philosophy – The Green Series: Toward a Sustainable Environment*, Ed. J. Newman, Vol. VIII, pp. 410–413, Sage Publisher, 2011.
17. F.R. Spellman, *Handbook of Water and Wastewater Treatment Plant Operations*, 3rd ed., CRC Press, 2013.
18. E.R. Alley, *Water Quality Control Handbook*, Vol. 2, McGraw-Hill, 2007.
19. S. Chidiac, P. El Najjar, N. Ouaini, Y. El Rayess, and D. AEl Azzi, *Rev. Environ. Sci. Biotechnol.* 22 (2023) 349–395, https://doi.org/10.1007/s11157-023-09650-7
20. S. Sivaranjani, A. Rakshit, and S. Singh, *Int. J. Bio. Sci.* 2 (2015) 85, https://doi.org/10.5958/2454-9541.2015.00003.1.
21. A. Chatterjee, *Water Supply Waste Disposal and Environmental Pollution Engineering (Including Odour, Noise and Air Pollution and Its Control)*, 7th ed., Khanna Publishers, 2001.

22. N.H. Omer, *Water Quality Parameters, Water Quality – Science, Assessments and Policy, Kevin Summers*, IntechOpen, 16 October 2019, https://doi.org/10.5772/intechopen.89657, https://www.intechopen.com/books/water-quality-science-assessments-and-policy/water-quality-parameters.

23. N.F. Gray, *Drinking Water Quality: Problems and Solutions*, 2nd ed., Cambridge University Press, 2008.

24. F.R. Spellman, *The Drinking Water Handbook*, 3rd ed., CRC Press, 2017.

25. APHA, *Standard Methods for the Examination of Water and Wastewater*, 21st ed., American Public Health Association, 2005.

26. M.L. Davis, *Water and Wastewater Engineering – Design Principles and Practice*, McGraw-Hill, 2010.

27. J.K. Edzwald, *Water Quality and Treatment a Handbook on Drinking Water*, McGraw-Hill, 2010.

28. P. Read and T. Fernandes, *Aquaculture* 226 (2003) 139–163.

29. W. Viessman and M.J. Hammer, *Water Supply and Pollution Control*, 7th ed., Pearson Prentice Hall, 2004.

30. S.H. Abbas, I.M. Ismail, T.M. Mostafa, and A.H. Sulaymon, *J. Chem. Sci. Technol.* 3 (2014) 74–102.

31. C. White, J. Sayer, and G. Gadd *FEMS Microbiol. Rev.* 20 (1997) 503–516.

32. G. Tchobanoglous, H.S. Peavy, and D.R. Rowe, *Environmental Engineering*, McGraw-Hill Interamericana, 1985.

33. M. Tomar, *Quality Assessment of Water and Wastewater*, CRC Press, 1999.

34. J. DeZuane, *Handbook of Drinking Water Quality*, 2nd ed., John Wiley and Sons, 1997.

35. G. Tchobanoglous, F.L. Burton, and H.D. Stensel, *Metcalf and Eddy Wastewater Engineering: Treatment and Reuse*, 4th ed., Tata McGraw-Hill Limited, 2003.

36. World Health Organization, *Guidelines for Drinking-Water Quality*, 4th ed., WHO, 2011.

37. M.L. Davis and A. David, *Introduction to Environmental Engineering*, 4th ed., McGraw-Hill, 2008.

38. T.J. McGhee and E.W. Steel, *Water Supply and Sewerage*, McGraw-Hill, 1991.

39. M. Natarajan, P. Raja, M. Gurusamy, and S. Rajagopal, *Curr. Res. J. Biol.* 1 (2009) 72–77.

40. Z. Hubálek, *Folia Zool.* 49 (2000) 241–260.

41. C.N. Sawyer, P.L. McCarty, and G.F. Parkin, *Chemistry for Environmental Engineering and Science*, 5th ed., McGraw Hill Education (India) Private Limited, 2003.

42. Y.-Y. Choi, S.-R. Baek, J.-I. Kim, J.-W. Choi, J. Hur, T.-U. Lee, C.-J. Park, and B.J. Lee, *Water* 9 (2017) 409, https://doi.org/10.3390/w9060409.

43. F.M. Wilhelm, Pollution of Aquatic Ecosystems I, In *Encyclopedia of Inland Waters*, Ed. G.E. Likens, pp. 110–119, Academic Press, 2009, https://doi.org/10.1016/B978-012370626-3.00222-2.

44. H. Leclerc, D.A.A. Mossel, S.C. Edberg, and C.B. Struijk, *Annu. Rev. Microbiol.* 55(1) (2001) 201–234.

45. O. Viman, I. Oroian, and A. Fleseriu, *Aquac. Aquar. Conserv. Legis.* 3 (2010).

46. N.N. Rabalais and R.E. Turner, *Limnol. Oceanogr. Bull.* 28 (2019) 117–124, https://doi.org/10.1002/lob.10351.

47. A. Evans, M. Hanjra, Y. Jiang, M. Qadir, and P. Drechsel, *Int. J. Water Resour. Dev.* 28 (2012) 195–216, https://doi.org/10.1080/07900627.2012.669520.

48. E.M. Mockler, J. Deakin, M. Archbold, L. Gill, D. Daly, and M. Bruen, *Sci. Tot. Environ.* 601–602 (2017) 326–339.

49. R.K. Sharma, M. Yadav, and R. Gupta, Chapter Five – Water Quality and Sustainability in India: Challenges and Opportunities, In *Chemistry and Water*, Ed. S. Ahuja, pp. 183–205, Elsevier, 2017.

50. R. Sharma, R. Kumar, S.C. Satapathy, N. Al-Ansari, K.K. Singh, R.P. Mahapatra, A.K. Agarwal, H.V. Le, and B.T. Pham, *Front. Environ. Sci.* 8 (2020) 581591.

51. S. Gangwar, *Int. J. Inf. Comp. Technol.* 3 (2013) 851–856.

52. R. Khaiwal, A. Meenakshi, M. Rani, and A. Kaushik, *J. Environ. Monitor (JEM)* 5 (2003) 419–426, https://doi.org/10.1039/B301723K.

53. D. Malik, S. Singh, J. Thakur, R.K. Singh, A. Kaur, and S. Nijhawan, *J. Curr. Microbiol. App. Sci.* 3 (2014) 856–863.

54. P. Kotoky and B. Sarma, *Int. J. Eng. Res. Technol.* 6 (2017) 536–540.

55. S. Dutta and S. Nayek, Water Quality of the Ganges and Brahmaputra Rivers: An Impact Assessment on Socioeconomic Lives at Ganga–Brahmaputra River Basin, In *Sustainability in Environmental Engineering and Science*, Ed. S. Kumar, A. Kalamdhad, and M. Ghangrekar, Lecture Notes in Civil Engineering, Vol. 93, Springer, 2021, https://doi.org/10.1007/978-981-15-6887-9_26.

56. C. Maharana, S. Gautam, A. Singh, and J. Tripathi, *J. Earth Syst. Sci.* 124 (2015) 1293–1309.

57. A. Das, T.K. Nath, and B. Tripathy, *Int. J. Eng. Res. Technol.* 7 (2018) 261–270.

58. S. Panigrahi and A. Patra, *Ind. J. Sci. Res.* 4 (2013) 211–217.

59. V. Kumar, S. Kumar, S. Srivastava, J. Singh, and P. Kumar, *Arch. Agri. Environ. Sci.* 3 (2018) 58–63.

60. R. Ramya Priya and L. Elango, *Environ. Earth Sci.* 77(2) (2018), https://doi.org/10.1007/s12665-017-7176-6.

61. G. Singh, N. Patel, T. Jindal, et al., *Environ. Monit. Assess.* 192 (2020) 394, https://doi.org/10.1007/s10661-020-08307-0.

62. D. Saksena, R.K. Garg, and R. Rao, *J. Environ. Bio.* 29 (2008) 701–710.

63. S. Makwana, *Int. J. Res. App. Sci. Eng. Technol.* 8 (2020) 21–28, https://doi.org/10.22214/ijraset.2020.4005.

64. S. Abidin, G. Rajendran, R. Ganapathi Raman, R. Selvaraju, and R. Valliappan, *Ind. J. Environ. Ecoplan.* 16 (2009) 193–198.

65. D. Gupta, R. Shukla, M. Barya, G. Singh, and V. Mishra, *Water Sci.* 34 (2020) 202–212.

66. D. Haldar, S. Halder, P. Das, and G. Halder, *Desal. Water Treat.* 57 (2016) 3489–3502.

67. A. Kshirsagar, and V.R. Gunale, *J. Ecophysiol. Occup. Health.* 11 (2011) 81–90.

68. P. Cheepi, IOSR *J. Environ. Sci. Toxicol. Food Technol.* 1 (2012) 40–51, https://doi.org/10.9790/2402-0144051.

69. 2006 World Health Organization, *Protecting Groundwater for Health: Managing the Quality of Drinking-water Sources*, Ed. O. Schmoll and G. Howard, WHO International Drinking Standards for Drinking Water, 3rd ed., WHO, 1971.

70. Council Directive 98/83/EC of 3 November 1998 on the Quality of Water Intended for Human Consumption, Annex I: Parameters and Parametric Values, Part B: Chemical Parameters, EUR-Lex. Retrieved 30 December 2019, https://eur-lex.europa.eu/legal-content/EN/TXT/PDF/?uri=CELEX:32020L2184.

71. Health Canada, *Guidelines for Canadian Drinking Water Quality – Summary Table. Water and Air Quality Bureau, Healthy Environments and Consumer Safety Branch*, Health Canada, 2020.

72. J. Cotruvo, V. Kimm, and A. Calvert, *Drinking Water: A Half Century of Progress*, EPA Alumni Association, 1 March 2016.

73. Indian Standard Drinking Water Specification (PDF), Central Ground Water Board, https://pcb.assam.gov.in/information-services/indian-standard-specifications-for-drinking-water.

74. K. Brindha and L. Elango, Fluoride in Groundwater: Causes, Implications and Mitigation Measures, In *Fluoride Properties, Applications and Environmental Management*, Ed. S.D. Monroy, pp. 111–136, 2011, https://www.novapublishers.com/catalog/product_info.php?products_id=15895.

75. M. Barathi, A. Santhana Krishna Kumar, and N. Rajesh, *Coord. Chem. Rev.* 387 (2019) 121–128.
76. S. Srivastava and S.J.S. Flora, *Curr. Environ. Health Rep.* 7 (2020), https://doi.org/10.1007/s40572-020-00270-9.
77. M. Singh Sankhla and R. Kumar, *ARC J. For. Sci.* 3 (2018) 10–15, https://doi.org/10.20431/2456–0049.0302002.
78. J.C. Barnett, *Stud. Conserv.* 52 (2007) 67–77.
79. A. Gürses, M. Açıkyıldız, K. Güneş, and M. Sadi Gürses, Dyes and Pigments: Their Structure and Properties, In *Dyes and Pigments*, pp. 13–29, Springer, 2016.
80. P. Gregory, Classification of Dyes by Chemical Structure, In *The Chemistry and Application of Dyes: Topics in Applied Chemistry*, Ed. D.R. Waring and G. Hallas, Springer, 1990, https://doi.org/10.1007/978-1-4684-7715-3_2.
81. K.K. Samanta, P. Pandit, P. Samanta, and S. Basak, Chapter 3 – Water Consumption in Textile Processing and Sustainable Approaches for Its Conservation, In *Water in Textiles and Fashion*, Ed. S.S. Muthu, pp. 41–59, Woodhead Publishing, 2019.
82. T. Robinson, G. Mcmullan, R. Marchant, and P. Nigam, *Bioresour. Technol.* 77 (2001) 247–255, https://doi.org/10.1016/S0960-8524(00)00080-8.
83. G. Sandin and G.M. Peters, *J. Cleaner Prod.* 184 (2018) 353–365.
84. P. Nigam, G. Armour, I. Banat, D. Singh, and R. Marchant, *Bioresource Technol.* 72 (2000) 219–226, https://doi.org/10.1016/S0960-8524(99)00123-6.
85. T.K. Chung, *J. Environ. Sci. Health, Part C* 34 (2016), 233–261, https://doi.org/10.1080/10590501.2016.1236602.
86. N.Z. Sekuljica, N.Z. Prlainovic, J.R. Jovanovic, A.B. Stefanovic, V.R. Djokic, D.Z. Mijin, and Z.D. Knezevic-Jugovic, *Bioproc. Biosys. Eng.* 39 (2016) 461–472.
87. S. Khan and A. Malik, *Environ. Deter. Human Health Natur. Anthropog. Determin.* (2013) 55–71, https://doi.org/10.1007/978-94-007-7890-0_4.
88. S.S. Muthu, Introduction, Chapter 1, In *Sustainability in the Textile Industry*, Ed. S.S. Muthu, pp. 1–8, Springer, 2017.
89. K. Hunger, *Industrial Dyes: Chemistry, Properties and Applications*, Willey-VCH, 2003.
90. H. Li, R. Zhang, L. Tang, J. Zhang, and Z. Mao, *J. Environ. Sci.* 26 (2014) 1125–1134.
91. I. Haq and A. Raj, *Chemosphere* 196 (2018) 58–68.
92. A.Y. Khan and G.S. Kumar, *Spectrochim. Acta A Mol. Biomol. Spectrosc.* 152 (2016) 417–425.
93. A. Petzer, B.H. Harvey, G. Wegener, and J.P. Petzer, *Toxicol. Appl. Pharmacol.* 258 (2012) 403–409.
94. G. Di Giovanni, V. Di Matteo, and E. Esposito (Eds.), *Serotonin-Dopamine Interaction: Experimental Evidence and Therapeutic Relevance*, p. 172, Elsevier, 2008.
95. N. Couto, J. Wood, and J. Barber, *Free Radic. Biol. Med.* 95 (2016) 27–42.
96. F.M.D. Chequer, J.P.F. Angeli, E.R.A. Ferraz, M.S. Tsuboy, J.C. Marcarini, M.S. Mantovani, et al., *Mutat. Res. Genet. Toxicol. Environ. Mutagen.* 676(1–2) (2009) 83–86.
97. E.A. Petrakis, L.R. Cagliani, P.A. Tarantilis, M.G. Polissiou, and R. Consonni, *Food Chem.* 217 (2017) 418–424.
98. C.V. Di Anibal, L.F. Marsal, M.P. Callao, and I. Ruisánchez, *Spectrochim. Acta A Mol. Biomol. Spectrosc.* 87 (2012) 135–141.
99. M. Piątkowska, P. Jedziniak, M. Olejnik, J. Żmudzki, and A. Posyniak, *Food Chem.* 239 (2018) 598–602.
100. K. Lacasse and W. Baumann, *Textile Chemicals: Environmental Data and Facts*, Springer, 2012.
101. S. Mani and R.N. Bharagava, *Reviews of Environmental Contamination and Toxicology*, Ed. P.de Voogt, Vol. 237, pp. 71–104, Springer, 2016.

102. C.F. Corvalán, H.N.B. Gopalan, and P. Llans, Conclusions and Recommendations for Action, Chapter 13, In *Climate Change and Human Health, Risks and Responses*, Ed. A.J. McMichael, D.H. Campbell-Lendrum, C.F. Corvalán, K.L. Ebi, A.K. Githeko, and J.D. Scheraga, WHO, WHO Library Cataloguing-in-Publication Data, 2003.

103. R.D. Moore, D. Spittlehouse, P. Whitfield, and K. Stahl, Weather and Climate, Draft, Chapter 3, In *Compendium of Forest Hydrology and Geomorphology in British Columbia*, pp. 3–55, B.C. Ministry of Forests and Range, Forest Science Program and FORREX Forum for Research and Extension in Natural Resources.

104. T. Wu, A. Hu, F. Gao, et al., *Clim. Atmos. Sci.* 2 (2019) 18, https://doi.org/10.1038/s41612-019-0075-7.

105. T.V. Ramachandra and A.V. Nagarathna, *Climate Change and Groundwater*. Ed. W. Dragoni and B.S. Sukhija, xiii + 186 pp, Geological Society, Special Publication 288. The Geological Society.

106. K. Abbass, M.Z. Qasim, H. Song, M. Murshed, H. Mahmood, and I. Younis, *Environ. Sci. Pollut. Res.* 29 (2022) 42539–42559, https://doi.org/10.1007/s11356-022-19718-6.

2 Separation process
Imperative to human life

2.1 MIXTURE AND SEPARATION

Matter is defined as anything that occupies space, has mass, and surrounds mankind. Man has always been associated with material from ancient times. The civilization periods have been named after the material used predominantly during that period (e.g. Stone, Bronze, and Iron). Silicon Age is refers to the modern period of history from the late 20th to the early 21st century. Matter can be classified into two broad categories as pure and mixture. A pure substance has a constant composition and property. It can be further divided into elements and compounds. Elements are substances which cannot be broken further into simpler substances, while compounds constitute of more than one element and can be broken further. Mixture is composed of two or more types of matter in varying amounts and can be separated using different techniques. The main difference between a mixture and a compound is that a compound is formed by chemical combination of elements of a particular state, while a mixture is usually a concoction of elements or even compounds of varying states of matter in different amounts. A mixture can be homogeneous (same state of matter) or heterogeneous (different states of matter). Mixtures form an essential part of man, and the need to separate it becomes important. Therefore, humans have always been associated with the process of separation for obtaining a pure material or removing unwanted material. In our daily life, the most common practices like cleaning dirt from one's hand or clothes, filtering of tea dust prior to drinking, etc., are carried out as part of the routine. Separation has always been part of human life as seen from various ancient texts. In ancient India, separation for water treatment has been mentioned in ancient texts. Sushruta demonstrated the use of nuts, gomedaka, lotus bulbs, moss, pearls, thick cloth, etc., for the removal of impurities, including those suspended in water (Sushruta, sutra 45.13) [1]. Some of the techniques that were used for purification were separation processes involving the addition of sand, mud, arjuna, musta, usira, nagakesara, kosataka, and amalaka, together with jetakaphala. Different techniques like pindavasa [lump of earth (alum), well mixed with phana, mustaka, ela, usira, and candana baked well in the fire of khadira and then dropped in water], puspavasa (with flowers), and curnadhivasa (powders) were adopted. The importance of water was also known to mankind and was always reminded to each other and the next generations in their daily talks and stories. The famous proverb in Tamil language of "Nīraiyumcīrāṭu", which literally means "Rehabilitate water", shows the importance of separation for purification of water. An anthology of the ancient Tamil period mentions the process of cleaning water using nuts (*Kalitokai*, Neithar Kali by Poet Nallanthuvanar) [2]. Traditional practices of placing a mud pot on sand and addition of herbs to water also indicate the use of separation processes in daily life. African countries used drumstick for water purification [2].

DOI: 10.1201/9781003442516-2

Separation is defined as a method by which a mixture or solution containing different components is resolved into individual components [3]. Thus, separation leads to enrichment of a particular phase with one of the components. It is seen that no separation is nearly complete and may have to be repeated [3]. Separation can be either analytical or preparative in nature and can be carried out at the lab scale or industrial level, respectively. In the analytical method, separation is utilized to segregate the required component with a high degree of purity to enable highly accurate and precise determination. Preparative separation aims at achieving a component with a high degree of purity for further applications.

Separation is a mass transfer process [4] which involves partitioning (distribution of species between different phases) due to differences in properties, namely, chemical (chemical affinity) and physical (size, shape, mass, and density) properties of species. During separation, one of the two phases gets considerably enriched with one component and the other phase with the second component of the starting mixture. The respective distributed components can be recovered from the two phases by using appropriate procedures. The different phase distributions lead to gas–gas, gas–liquid, gas–solid, liquid–liquid, liquid–solid, and solid–solid separations with large applications. Each of these phases leads to separation with different applications. These applications all contradict the old Latin axiom known to chemists from ancient times: "*Corpora non agunt nisi fluida (or liquida) seusoluta*", which simply means "Compounds do not react unless fluid or if dissolved" [5]. However, it was later thought that there could be some errors in the interpretation of the original text, and it should possibly mean "it is chiefly the liquid substances which 'react'" or "for instance, liquids are the type of bodies most liable to mixing" [5]. It is quite understandable that solid-state chemists could not agree with the first version of Aristotle's statement. A large number of separation processes are available, and the appropriate method should be chosen based on the application.

2.2 CLASSIFICATION OF SEPARATION TECHNIQUES

A large number of separation techniques are available. These techniques can be grouped together under a specific classification to achieve a better understanding and make a correct choice of the technique. There are many ways to classify the techniques [6].

The first mode of classification is based on the nature of the mixture. In this mode, the solute can be either a solid or a liquid dissolved in a solvent. These techniques can be further divided into homogeneous and heterogeneous mixtures. If the solid is completely dissolved in the solvent and only one phase is seen, then it is known as a homogeneous mixture, e.g. saline solution. Such mixtures can be separated using evaporation. A heterogeneous solution of solid in liquid contains the solid in a dispersed or colloidal form, e.g. muddy water. Sedimentation, filtration, and centrifugation are some of the techniques which can be grouped into this category. When the solute is in a liquid form and is completely miscible with the solvent, e.g. alcohol–water homogeneous mixtures, the process of distillation is most appropriate. However, if the liquid solute phase is not miscible with the matrix solvent, then it forms heterogeneous liquid-in-liquid mixtures, e.g. oil in water mixture, which can be separated using partition techniques.

Separation techniques can also be classified based on the property of the solute that is to be separated. The property used can be either physical or chemical in nature. Some of the processes dependent on the physical property of the solute are crystallization, centrifugation, decantation, filtration, sieving, and magnetic separation.

Centrifugation is the modus operandi and is based on the use of centrifugal force for the separation of particles from a solution based on physical properties (size, shape, density, viscosity of the medium, and rotor speed), with the denser components migrating away from the centre. The application of large effective gravitational pull, as in a normal centrifuge, leads to faster and efficient separation of particles. Crystallization (natural or artificial) involves nucleation (appearance of a crystalline phase from either a super cooled liquid/supersaturated solvent) and crystal growth (increase in particle size) and is used to separate mixtures of salts or even covalent solids that have different solubilities in a solvent. It is a solid–liquid separation technique involving mass transfer from a liquid to a pure solid crystalline phase. It is used for the purification of salt obtained from seawater. Size exclusion chromatography (SEC) and gas chromatography (GC) are based on the physical properties of size or physical sorption ability, respectively. SEC, also known as molecular sieve chromatography, leads to the separation of molecules in a solution based on their size or molecular weight. It is usually applied to proteins and industrial polymers. Inverse gas chromatography (IGC) is a physical characterization analytical technique used to evaluate the surface and bulk properties of solids. In IGC, the roles of the stationary (solid) and mobile (gas or vapour) phases are inverted from that of traditional analytical GC. Whereas in GC, a standard column is used to separate and analyse gases, in IGC, a single gas or vapour (probe molecule) is injected into a column packed with the solid sample under investigation, and the retention time of the probe gas molecule is measured with traditional GC detectors (i.e. flame ionization detector or thermal conductivity detector). Decantation is a separation process used for mixtures of immiscible phases (two immiscible liquids or a solid suspension or settled in liquid), where the less dense layer is just carefully poured off, leaving the other component behind. This is not quite effective for separation of two immiscible liquids. Demister is a device used to remove liquid droplets from a vapour stream by reducing the residence time required to separate and is used for applications wherein high vapour quality is crucial. Drying is a mass transfer process consisting of the removal of water or another solvent by evaporation from a solid using a heat source and a means to remove the vapour produced. Electrophoresis is the separation of charged species under the influence of an applied electric field, based on preferential migration. Gel electrophoresis is used to separate macromolecules (DNA, RNA, and proteins) based on size and/or charge. Elutriation is a separation method (based on size, shape, and density) for particles smaller than 1 μm using a stream of flowing gas or liquid (in a direction usually opposite to the direction of sedimentation). Evaporation is a type of vaporization on the surface of a liquid as it changes into the gas phase and then reaches an equilibrium. Froth flotation is the process of gathering hydrophobic materials from hydrophilic materials in and on the surface of a froth layer, leading to selective separation. Fractional distillation is the separation of a mixture into its components by heating to a temperature at which one or more fractions of the mixture will vaporize and is distilled. Fractional freezing is the oft used method in process engineering and chemistry for the separation of substances with different melting points by partial melting of a solid. Magnetic separation is the process of separating components of mixtures by using magnets to attract magnetic materials, thus separating them from non-magnetic material. Porous neutral membranes used in processes like ultrafiltration, microfiltration, and nanofiltration bring out separation based on size. Recrystallization is a technique used to purify chemicals by dissolving a compound in an appropriate solvent and remove either the desired compound or impurities from the solution. Sedimentation is a process based on the motion of particles through fluid before they settle against a barrier, leading to easy separation between solid and solution phases. The motion of the suspended particles through the fluid can be an effect of the forces (gravity, centrifugal acceleration, or electromagnetism) acting on them. If the components have different specific weights, gravitational force becomes important, and the process is known as gravity separation. Sieving is the separation of particles based on the difference in sizes of the particles using sieves of different sizes. The design of the industrial sieve is of primary importance. This separation finds an important role in food industries wherein vibrating sieves are used to prevent the contamination. Sublimation is an endothermic changeover of a substance from the solid to gas state directly at temperatures and pressures below a substance's triple point, and its reverse is deposition or desublimation (a substance passes directly from a gas to a solid phase). In recent times, the term sublimation is used to construe both solid-to-gas and gas-to-solid transitions. It is to be understood that the change here refers to a change in the physical state of the compound, and not a chemical change, caused by the absorption of heat by molecules and conversion to the vapour phase. Separation due to sublimation is possible because each compound has its own unique sublimation temperature. Wind winnowing is a separation procedure used from time immemorial for applications in agriculture. It involves the flinging of a mixture into an air current so that heavier grains are retained while lighter impurities are wafted away by the wind. The different methodologies adopt various parts like a winnowing fan (a specially shaped basket shaken to raise the chaff) or use a tool (a winnowing fork or a shovel) to facilitate separation. Zone melting (zone refining or floating-zone process or travelling melting zone) is a group of similar methods used to decontaminate crystals by moving the impure melt containing impurities forward, leaving behind the pure material.

All the above methods (based on differences in the physical properties of the solute) can be very selective and complete only if the differences in these properties are very large. In order to make the separations more efficient and selective, differences in the chemical properties of the solute have to be exploited. Some of the processes are discussed. Ion exchange and chromatographic separations are mass transfer sorption processes involving selective transfer of sorbates from the fluid phase to the surface of the insoluble, rigid solid phase. The differences in partition coefficients result in differential retention on the stationary phase, and thus separation is achieved. Chromatography can be preparative or analytical. Preparative chromatography separates the components of a mixture which can be used in purified form for some other applications. Analytical chromatography is carried out with smaller quantities and can be used for the determination and identification of substances. High-performance liquid chromatography (HPLC; formerly known as high-pressure liquid chromatography) is a technique in analytical chemistry used to assess the nature and quantity of components of a mixture due to their differences in the interaction with both the solid and liquid phases. HPLC is different from traditional (low-pressure) liquid chromatography in that the operational pressures are significantly higher in HPLC, while ordinary liquid chromatography typically relies on the force of gravity to let the mobile phase pass through the column. Thin layer chromatography (TLC) is a technique used to separate non-volatile mixtures on a sheet of glass, plastic, or aluminium foil, coated with a thin layer of adsorbent material, usually silica gel, aluminium oxide (alumina), or cellulose, known as the stationary phase. The stationary phase with the sample is contacted with a solvent or solvent mixture (mobile phase), which goes up due to capillary action and separation is achieved. It can be used to monitor the progress of a reaction, identify compounds present in a given mixture, and determine the purity of a substance. Counter-current chromatography (CCC) is a form of liquid–liquid chromatography that uses a liquid stationary phase which is held in place by centrifugal force and used to separate, identify, and quantify the chemical components of a mixture. The resulting dynamic mixing and settling action allows the components to be separated by their respective solubility in the two phases. In droplet counter-current chromatography (DCCC), the mobile phase passes through the columns in the form of droplets. An excellent analytical method for the determination of coloured substances is paper chromatography. Two-dimensional paper chromatography is very useful for separating complex mixtures of compounds with comparable polarity, such as amino acids. The mobile phase (a mixture of non-polar organic solvent) traverses up the stationary phase (polar water held inside the void space of the cellulose network of the paper support) due to capillary action. This technique contrasts with TLC as the stationary phase in TLC is a layer of sorbent (usually silica gel or aluminium oxide). An excellent technique for the separation of polar components and inorganic ions is ion chromatography (or ion-exchange chromatography). This method is based on the affinity of these ions towards the ion exchanger. There are two types of ion chromatography, namely, cation-exchange and anion-exchange that are used for the separation of cations and anions, respectively. When cations are separated, the stationary phase is negatively charged, while for anions the stationary phase is positively charged. It is often used in water analysis and quality control. Affinity chromatography, used extensively for biochemical applications, works on the principle of a highly specific interaction (hydrogen bonding, ionic interaction, disulphide bridges, and hydrophobic interaction) between an antigen and an antibody. It is used for purifying biological molecules in a mixture in a laboratory. Centrifugal partition chromatography is a special chromatographic technique where both stationary and mobile phases are liquid, and the stationary phase is immobilized by a strong centrifugal force. Centrifugal partition chromatography consists of a series-connected network of extraction cells, which operate as elemental extractors, and the efficiency is assured by the deluge. Sorption is the adherence of atoms, ions, or molecules of a gas or liquid on to a surface, resulting in a layer of the adsorbate on the surface of the adsorbent. This is in contrast to absorption, in which the absorbate permeates into the bulk of the absorbent. Nowadays both these process are collectively known as sorption. Solid-phase extraction (SPE) is an extractive technique used to concentrate and purify samples for analysis. The result is that either the desired analytes of interest or undesired impurities in the sample are retained on the stationary phase. A great number of materials can be used as sorbents. Extraction is a separation process used for the separation of a substance from a matrix and is usually carried out using equilibria involving a liquid–liquid or a solid phase. The distribution of a solute between two phases is an equilibrium condition described by partition theory. Liquid–liquid extraction (LLE), also known as solvent extraction, separates compounds or metal complexes based on their relative solubilities in two different immiscible liquids, usually polar (water) and non-polar (organic solvent). There is a net transfer of one or more species generally from an aqueous to an organic solution, and the solvent enriched in solute(s) is the extract, while the feed solution depleted in solute(s) is known as a raffinate. LLE is a basic laboratory separation technique. Supercritical fluid extraction (SFE) is the process of separating one component (extractant) from another (the matrix) using supercritical fluids as the extracting solvent. SFE can be used as a sample preparation step for analytical purposes, or on a larger scale to either strip unwanted material from a product (e.g. decaffeination) or collect a desired product (e.g. essential oils). Carbon dioxide $\left(CO_2\right)$ is the commonly employed supercritical fluid. Precipitation is the production of a solid, known as precipitate, from a solution due to a reaction with the reagent known as the precipitant. The precipitate-free liquid remaining above the solid is called the "supernate" or "supernatant". Precipitation occurs when the concentration of a compound exceeds its solubility product; therefore, the kinetics of precipitation is very high from a supersaturated solution.

Another mode of classification is that based on the mechanism followed. There are three mechanisms, namely chemical equilibrium, kinetic, and mechanical, which control the separation efficiency. The first is based on chemical equilibrium that is achieved, and it depends on the different phases involved in separation. Distillation is a process based on the equilibrium between the vapour and liquid phases, while both sublimation and sorption involve solid and vapour phases. Sorption uses ion exchange and charged membranes and sorbents, and sublimation is based on the chemical equilibrium between a solid and a liquid, while LLE involves two immiscible liquid phases. The classification based on mechanical phase separation involves procedures like precipitation, filtration, and extraction. It is seen that classification based on kinetics would probably encompass all the processes. It is the difference in the kinetics of a particular procedure which leads to separation irrespective of the property of the solute, equilibrium, or mechanical phases involved. Ion exchange membranes bring about selectivity due to their inherent property of permselectivity, which refers to the difference in the permeability of ions of opposite charges, i.e. a cation exchange membrane will not allow the transport of anions.

2.3 CONCLUSIONS

The separation process which can be used for treatment depends on various factors like wastewater characteristics [7]. The processes are associated with their own merits and demerits assessed from various aspects like cost, feasibility, efficiency, practicability, reliability, environmental impact, secondary waste generation, etc. The development of cheaper, effective, and novel methods of decontamination is an ongoing research. It is also difficult to define a universal method that could be used for the removal of all pollutants from waste waters.

In the ensuing chapters, we will look into some of the commonly used techniques for the removal of metal ions and cationic dyes from waste water. In these chapters, the discussion will be focussed on the basic principles followed by some examples of applications. The readers will surely appreciate the fact that since most of these methods are continuously developing, a large number of systems are available in the literature. Since it is not practically possible to cite all these works, some of these studies have been given as representative examples.

REFERENCES

1. R. Krishnamoorthy, *Ind. J. His. Sci.* 34 (1996) 327–337.
2. S.L. Bansal and S. Asthana, *Ind. J. App. Res.* 9 (2019) 16–18.
3. I.D. Wilson, E.R. Adlard, M. Cooke, et al. (Eds.), *Encyclopedia of Separation Science*, Academic Press, 2000, ISBN 978-0-12-226770-3.
4. D. Basmadjian, *Mass Transfer and Separation Processes: Principles and Applications*, 2nd ed., CRC Press, https://doi.org/10.1201/9781420051605.
5. C. Reichardt, *Org. Proc. Res. Dev.* 11 (2007) 105–113.
6. L.N. Moskvin, *Sep. Purif. Rev.* 45 (2016) 1–27.
7. Y. Anjaneyulu, N. Sreedhara Chary, and D. Samuel Suman Raj, *Rev. Environ. Sci. Bio. Technol.* 4 (2005) 245–273, https://doi.org/10.1007/s11157-005-1246-z.

3 Membrane separation
Primordial water treatment technique

3.1 WHAT IS A MEMBRANE?

The very first mention of membrane-based separation was in the mid-18th century. In 1748, Jean-Antoine Nollet, also known as Abbe Nollet, conducted an experiment. He took a vial of alcohol (with air purged out), covered it securely with a pig's bladder, and then submerged it into a container of water. After 6 h, he noticed that a piece of the pig's bladder had bulged [1, 2]. However over years, membranes have become an important aspect of separation used in different applications. Membrane separation is a process in which different components in a mixture are separated using a membrane [3]. Separation is achieved by the retention of unwanted species and allowing permeation (transport) of species of interest. It is also possible to achieve separation at different permeation rates. Several factors are found to affect the separation efficiency. The size, polarity, and ionic charge of the solute; physical and chemical characteristics of the membrane; and external parameters like temperature and pH of the solution are found to have a strong effect on permeation rates and selectivity.

According to the IUPAC nomenclature, a membrane is defined as a structure having lateral dimensions much greater than its thickness, through which transfer may occur under a variety of driving forces [4]. The barrier can result in the transfer of various kinds of components in the liquid or vapour phase. Separation is achieved if one component travels faster than the rest of them in the mixture. Separation using membrane is schematically represented in Figure 3.1 [5].

A semipermeable (also known as partially or differentially permeable) membrane is a barrier that will only allow some molecules to pass through (depending on the attributes of the molecules such as size) while blocking the passage of other molecules and acting as a filter. The passage of components occurs due to diffusion from a region of high concentration (of a component) to a region of low concentration. The membrane can be either natural (biological) or synthetic in nature. A biological example of a semipermeable membrane is the kidney tissue, which allows only certain molecules to pass through while blocking others. Synthetic versions of a semipermeable membrane are polymers used for water filtration or desalination applications. There are different types of permeable membranes. A permeable membrane allows the passage of all materials through it, while an impermeable membrane does not allow the passage of materials through it. A semipermeable membrane allows some materials to pass through it based on size, while a selectively permeable membrane allows only specific components to pass through. Generally, a semipermeable membrane does not allow solutes to through pass through it, and only one type of solvent

DOI: 10.1201/9781003442516-3

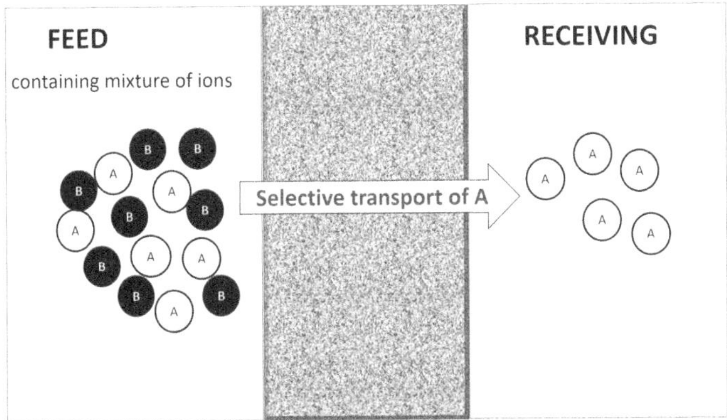

FIGURE 3.1   Schematic representation of separation across a membrane

flows through it. It acts as an ideal partition between two osmotically active solutions. A selectively permeable membrane allows selected solutes and solvents to pass through it and acts as an imperfect partition. It permits the entry of both solvents and, to a selected extent, solutes.

3.2   CLASSIFICATION OF MEMBRANES

Membranes can be either natural or synthetic in nature. However, in this chapter, the discussion is restricted to synthetic membranes, which are used for the separation of toxic species. Membranes are broadly classified into solid and liquid membranes (LMs) [6].

3.2.1   SOLID MEMBRANES

Synthetic solid membranes can be classified based on their nature into organic or inorganic membranes [7]. The membrane material can be either natural or synthetic. Synthetic membranes can be organic or inorganic in nature. Membrane structures can be anisotropic or isotropic. Composites and membranes synthesized by the phase separation method can give an anisotropic structure. Mesoporous, non-porous, and electrically charged membranes have an isotropic structure.

Ceramic inorganic membranes (produced from inorganic oxides, glass, zeolites, and carbide) are found to be possess excellent chemical and thermal stability. Organic membranes are polymers (derived from the Greek words *poly* and *mere*, meaning many and part, respectively) composed of many repeating units of an identical structure. Polymers can be natural (protein, cellulose, and silk) or synthetic (polystyrene, polyethylene, and nylon). Ceramic membranes and polymeric membranes have different properties. Ceramic membranes have uniform pores unlike polymeric membranes, thus allowing the sieving mechanism to operate. Polymeric membranes have a tendency to swell when in contact with aqueous solutions unlike ceramic membranes. Ceramic membranes are chemically resistant and thermally stable compared to polymeric membranes. However, the main disadvantages are that ceramic membranes are more brittle and very expensive compared to polymeric membranes. Synthetic membranes can be classified based on the charge associated with them into neutral and electrically charged membranes. Neutral membranes do not have any functional groups which can be used for the exchange of cations, and separation with these neutral membranes is based on the sieving mechanism. Neutral membranes can be classified based on their structure (morphology) into symmetric or asymmetric membranes. Symmetric membranes (porous and non-porous) have thickness in the range of 10–20 μm, while asymmetric membranes contain a very dense top layer (0.1–0.5μm thick) supported by a porous sub-layer (50–150 μm thick). Symmetric membranes can be porous or non-porous in nature. The non-porous membranes may contain a very thin layer of a dense material and find great use for gas separations and reverse osmosis (RO) applications. Dense membranes can be either amorphous or heterogeneous in nature. Porous membranes are used in different membrane filtration processes. The most commonly used theory for defining a porous membrane assumes that the membrane structure consists of parallel, non-intersecting cylindrical capillaries. However, actually, it may be a random network of different sizes which are affected by synthetic conditions. The sizes of the pores give rise to various membrane filtration processes, and the applications also vary. It is seen that the membrane filtration processes are pressure driven in nature, and depending on the nature of the membranes, the pressure applied will also vary. Microfiltration (MF) membranes have pores that retain solutes of size ranging from 0.1 to 20mm and operate at 20–100 kPa. The process is based on sieving mechanism and has replaced the conventional sedimentation process. It is used for the retention of algae, animalcules, and bacteria. Ultrafiltration (UF) process uses 0.01-μm porous membranes and operates at 100–1,000 kPa. The main mechanism is sieving, and it has replaced the conventional centrifugation process. It is found to be suitable for the retention of solutes of size 5–100 nm and can be used for the removal of small colloids and viruses. Nanofiltration (NF) process using nanoporous membranes and an operating pressure of 500–1,500 kPa is used to retain solutes of size in the range of 0.5–5nm and so can retain dissolved oxygen matter (DOM) and divalent ions. The main operating mechanism is that of dissolution and diffusion, and it has been found to replace the conventional distillation and evaporation techniques in various applications. RO uses non-porous membranes and retains solutes like monovalent ions in the size range of 0.1–1 nm. The operating pressure is quite high, and the mechanism is based on dissolution and diffusion. RO has been useful in replacing the conventional distillation and evaporation techniques. Synthetic ion-exchange membranes (with groups similar to ion-exchange resins) are known as ionomers or polyelectrolytes depending on the ionic content. Ionomers have an ionic content of about 10–15 mol.%, and this is much higher in polyelectrolytes, making them water-soluble. The incompatibility of ionic aggregates (multiplets) and non-ionic segments (clusters) results in microphase separation in ionic polymers [8]. Ion pairs within multiplets induce temporary ionic cross-links, while large clusters result in mechanical reinforcement. Thus, the properties of these materials are very different from those of other polymers; thus, applications are numerous and diversified, including permselective permeation of ions and encapsulation. Perfluoro sulphonate ionomers (PFSIs) are copolymers of tetrafluoroethylene and perfluorinated vinyl ether containing a terminal sulphonyl fluoride (SO_2F) group. Walther Grot of DuPont de Nemours in the late 1960s developed a perfluoro sulphonate membrane, known as Nafion, with a polyfluoroethylene backbone and regularly spaced perfluorvinyl ether pendant side chains terminated by a sulphonate ionic group. Nafion membranes contain incompatible components: the fluorocarbon phase and the ionic phase (containing ions and water molecules). These are separated to the limit that the covalent bonds hold them together. There are two kinds of arrangements visualized for Nafion structure, namely, the cluster-network and the three-phase model. The cluster model proposes that the aqueous ions are embedded in a continuous fluorocarbon phase, and the clusters thus formed are interconnected by narrow channels (to determine the transport properties of ions and water molecules). The water content of the membrane is of great importance for its properties and is determined by the kind of polymers and counterions. According to the three-phase model, there is an intermediate region which contains an ether side chain, a fluorocarbon backbone, and also sulphonate groups and associated counterions [8]. The equivalent weight (EW) of the ionomer is given by the relationship $EW = 100*m + 446$, wherein m is the number of structural units. Therefore, for a membrane of 1,100 EW, the side chains have 14 CF_2 units. The designation "117" refers to a film having 1,100 EW and a nominal thickness of 0.007 inch.

Solid membranes can also be classified based on the geometry of the membrane used into tubular, hollow fibre, spiral wound, and flat sheet modules. Tubular membranes are not self-supporting but are located within tubes. Tubular membranes have a diameter of about 5–15 mm. Because of the size of the membrane surface, plugging of tubular membranes is not likely to occur. A drawback of tubular membranes is that the packing density is low. In the hollow fibre module, many of the fibres (each in the tubular module) are placed in a large pipe in a compact module. Hollow fibre membranes can be designed for dead-end circulation. The advantages of such modules are compact size, reduced cost, and high quality of water produced. In spiral wound membrane, the membrane is cast as a film onto a flat sheet. Membranes are sandwiched together with feed spacers (typical thickness 0.03–0.1 inch) and a permeate carrier. They are sealed at each edge and wound up around a perforated tube. The module has a diameter of 2.5–18 inch and a length of 30–60 inch. The hollow fibre module with the lowest manufacturing cost and highest packing density is very easily susceptible to fouling, and cleaning is very difficult [9] compared to other configurations. This module is used for dialysis, reverse osmosis, and ultrafiltration methods. The tubular wound module suited for RO and UF processes has the least packing density and is also quite expensive. But it has very less tendency to be fouled and can be easily cleaned.

Solid membranes have many advantages, which makes them very useful for a variety of applications. The energy requirement of solid membrane separations is low, making the design quite simple and very easy to operate. The process is made very selective by using novel membranes, making it quite desirable. The availability of a large number of organic polymers and inorganic materials makes it easy to achieve selective separation with great ease. These membranes can be used to remove trace-level components at low energy requirements. These processes are environment-friendly due to the use of simple and non-toxic materials. Despite the various advantages, solid membranes have some limitations. The separation can never produce a completely pure product in a stream but can be only concentrated as a retentate. The number of stages in membrane separation is usually one or at the maximum two, thus making it essential for the membrane to have high selectivity. The membranes used may show chemical incompatibility with process solutions, leading to dissolution, swelling, and weakening of the membrane structures. This reduces the capacity and shortens the membrane lifetime. The temperature of an operation cannot be very high as the membrane structure will be greatly affected. Scaling up of the membrane process to accommodate very large stream sizes becomes difficult as there will be a substantial increase in the number of required membrane modules. Another major associated problem is membrane fouling, which hampers the permeation rate and affects the separation efficiency. In order to circumvent the disadvantages of low permeation flux and decreased selectivity associated with solid membranes, the use of liquid membranes is an attractive alternative.

3.2.2 Liquid membranes

Liquid membranes have several other names, such as liquid pertraction, carrier-mediated extraction, facilitated transport, and two-stage extraction. The term liquid membrane defies the conventional image of a membrane. However, the term liquid pertraction [10] is useful to understand the process very clearly. It echoes a three-phase system containing two homogeneous aqueous solutions, referred to as feed (donor) and receiving (acceptor) solution, spatially separated by a third immiscible liquid, denoted as the membrane phase (M). The favourable thermodynamics at the feed–membrane interphase facilitates the extraction of solutes of interest into the membrane phase through which it is transported to the membrane-receiving interphase. Here it is stripped into the receiving phase using conditions different from that of extraction. Hence, liquid pertraction can be visualized to be a combination of solvent extraction and stripping processes in both time and space. This combination is advantageous over conventional extraction as it offers the maximum driving force. Moreover, the extraction capability of organic media is not crucial, making it possible to use inert and non-toxic organic liquids. In order to enhance extraction, small amounts of carriers, which are organic complexing reagents, can be added to the membrane liquid. The carrier forms a complex with the ions of interest, and the metal complex is extracted into the organic phase. The use of carriers can improve the selectivity of the membrane separation process. The chemical potential difference between the two aqueous solutions results in the transfer of mass, whereas the extent of solute diffusion is controlled by concentration difference. Liquid membrane can generally be classified as bulk liquid membrane (BLM), supported liquid membrane (SLM), and emulsion liquid membrane (ELM) based on configurations and the setup of feed (F), membrane (M), and receiving (R) phases [5,10, 11]. A schematic representation of the three membrane setups is presented in Figure 3.2 [5,10, 11].

BLM involves a large volume of the membrane phase spatially separating the F and R phases. In the SLM, the membrane phase is immobilized in the pores of a polymeric support, and this separates the two aqueous phases, namely, F and R. In the ELM, small globules of the M phase containing droplets of the R phase are dispersed

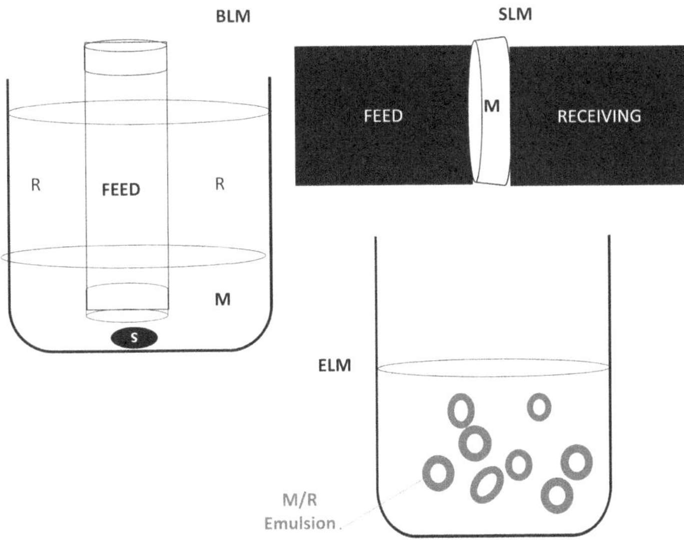

FIGURE 3.2 Schematic representation of different liquid membranes (drawn based on references in the literature [5,10, 11])

continuously in the F phase. The low solvent inventory of the SLM and the high flux of the ELM make these techniques attractive, but poor membrane stability hampers their extensive use. SLM is associated with the leaching out of the membrane layer, out of the pores of the polymeric membrane, due to the wetting and osmotic pressure across the membrane, while ELM is associated with the emulsion formulation method. Progress in membrane contactors using liquid membranes with integrated features of the SLM or ELM [hollow fibre SLM (HFSLM), pseudoemulsion-based hollow-fibre strip dispersion, or hollow fibre renewal liquid membrane] has still not been able to resolve the concern of membrane stability. However, BLM has no such issues as the problems of membrane rupture or exhaustion of carriers do not arise. The added advantages of BLM, such as its simplicity, cost-effectiveness, easy manoeuvrability, constant interfacial area, and hydrodynamic conditions, make it an excellent tool for the evaluation of carrier efficiency and reaction kinetics and mechanisms in the laboratory. BLM can be used in different setups based on the relative density of the membrane liquid compared to water [11]. The schematic representation of these configurations is shown in Figure 3.3.

The reduced solute flux (due to low interfacial area/volume) and slow kinetics (due to longer transport path and higher membrane resistance) limit the industrial applications of BLM. The different configurations of BLM affect the interfacial area and transportation path, while membrane resistance is influenced by M-phase viscosity, stirring speed, and operating temperature. Although BLM has superior stability over SLM and ELM, BLM cannot be scaled up for industrial applications due to its high membrane resistance resulting from the bulky membrane phase, thus hampering solute transport and affecting BLM performance. Various factors like M-phase viscosity, stirring speed, and operating temperature can have an effect on membrane phase resistance. The membrane phase is made up of a carrier, diluent,

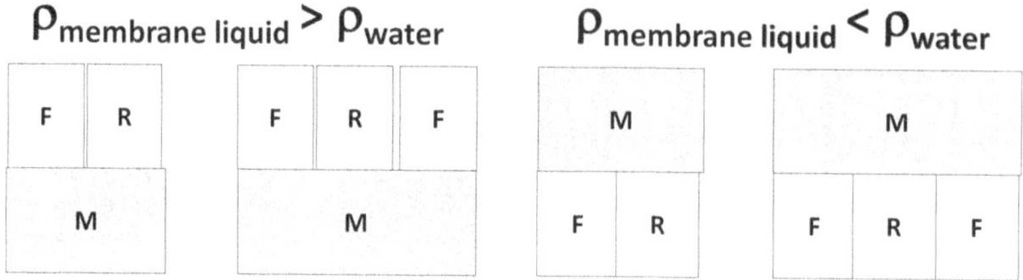

F and R are Feed and Receiving aqueous phases;
M is the organic membrane phase

FIGURE 3.3   Schematic representation of different configurations of BLM

and modifier. Solvents like organophosphorus compounds, amines, oximes, quinolines, and crown ethers are used in low levels $(<10\%)$ as carriers, while diluents constitute more than 85% of volume of the membrane phase, and solvents like alkanes, cycloalkanes, chlorinated alkanes, acyclic and cyclic ketones, alcohols, and aromatic compounds can be used. Sometimes solvents like alcohols, organophosphorus compounds, etc., are used in low concentrations $(<5\%/vol)$ as modifiers [12]. The carrier binds with the ion of interest and transports it from the feed to the receiving phase, while the diluent alters the metal ion hydration by increasing hydrophobicity and synergistically enhances the carrier activity. A modifier is normally used in cases where the problem of third-phase formation is severe. Volatile solvents are not preferred due to the instability of the liquid membrane and their quicker evaporation alongside environmental pollution. Low viscous, non-volatile solvents are ideal for LM applications. All these solvents have different viscosities attributable to the divergences in the intermolecular forces and shapes, which contribute to different extents of molecular resistance of the BLM. The viscosities are measured using water as a reference. It is seen that non-polar solvents are less viscous than polar solvents; the lowest and highest values of viscosity are 1.30 and 21.22 MPa*s for cyclohexane and di-2-ethylhexyl phosphoric acid, respectively, at room temperature [13]. Amongst the non-polar solvents, the least and most viscous solvents are hexane and kerosene with viscosity values of 0.30 and 1.64 MPa*s, respectively, at room temperature. Amongst polar solvents, the least and most viscous solvents are cyclohexanone and di-2-ethylhexyl phosphoric acid with values of 2.02 and 21.22 MPa*s, respectively, at room temperature [13]. The low viscosity of the non-polar solvents is attributed to the zero or quite small dipole moment values, resulting in a weak bond formation (van der Waals) between solvent molecules, which allows the molecules to easily slide over each other. The increased molecular size leads to increased van der Waals forces, thus making long-chain non-polar solvents (e.g. dodecane and kerosene) viscous. The low viscosity of water despite its high polarity and the presence of strong hydrogen bonds is due to its small molecular size. Therefore, the correct choice of membrane solvent helps in reducing the membrane resistance, leading to an increased transport rate with a high diffusion coefficient. Evaluation of the effect of a solvent on transport flux showed a substantial increase in flux values when the diluent was changed from viscous carbon tetrachloride to the less viscous chloroform or dichloromethane. This shows that a reduction in viscosity could increase the flux values [13]. The viscous resistance can be overcome by continuous stirring of the membrane phase as it reduces the viscous resistance and improves the hydrodynamics. Thus, the solute is transported at a faster rate due to convection as a result of stirring than by molecular diffusion in the unstirred state. The stirring exerts fluid forces on the solutes, which are physically moved from one spot to another by the random motion of small masses of fluid. The greater stirring speed results in higher fluid velocity, leading to enhanced solute transport. However, the increase in stirring speed does not increase the flux values as substantially as observed in the case of viscosity changes. There are reports on the distortion of feed/membrane and membrane/receiving interphases and also mixing of the two aqueous phases. The increase in operating temperature can also reduce viscous resistance in the membrane phase due to the decrease in cohesive forces between molecules, leading to an increase in kinetic energy of solvent molecules [13]. This reduces the fluid shear stress, and the solvent molecules become less viscous, leading to a substantial increase in the flux values. The effect of the increase in temperature on flux is more than that of stirring but still less than the effect of membrane solvent viscosity. The activation energy calculated indicated a diffusion-controlled, and not a chemical-controlled, reaction. The efficiency of a solvent is characterized by its distribution coefficient. Distribution coefficient is defined as the ratio of the amount of target solute present in the organic phase and the aqueous phase after equilibrium is reached [13]. A very high distribution coefficient indicates excellent extraction and vice versa. A strong extractant will not release the solute easily at the strip–membrane interphase. Therefore, thorough investigation is necessary to identify a suitable solvent–carrier combination that would give the best possible separation for a target solute. In this transport mechanism, the transported substance dissolves in the organic membrane phase and diffuses to the receiving phase due to the concentration gradient between these two phases. The selectivity of separation and/or the pertraction capacity can be increased if the permeant undergoes a chemical reaction in the stripping phase. The separation effectiveness is significantly improved if the component to be removed is transformed almost quantitatively into an impermeable form in the receiving phase. In this way, the concentration gradient can be maintained across the membrane, hence facilitating solute enrichment. Also, selectivity can be further enhanced by proper choice of the reaction window in the stripping solution so that only the compounds of interest are ionized and trapped.

Although BLM can remove metal ions effectively, it cannot be scaled up due to the toxicity and non-biodegradability of the solvents used. To reduce the toxicity, ionic liquids or vegetable oils are used to formulate the membrane phase [14]. Although the removal efficiency is quite comparable to that of the petroleum-based solvents already in use, the high viscosity hampers the kinetics. Ionic liquids are more advantageous than vegetable oils as their molecular structure can be easily modified, thus making it possible to tune their viscosity. However, their high cost, compared to that of vegetable oils, and probable toxic effects pose some reservation on its application on a large scale. To overcome all these disadvantages, new designs of BLM with features of spiral wound SLM and HFSLM and ELM are examined. The driving forces for improving the fluxes such as the electrical potential difference applied across a

BLM, by transporting metal ions between the phases and reducing solvent viscosity, make it possible to use safe but viscous solvents like oils for the membrane phase.

SLM is a non-dispersive-type LM, formed by the immobilization of a thin layer of an organic liquid phase (with dissolved reagents) onto a suitable inert microporous hydrophobic polymer support which does not play an active role in separation [15]. Based on the size, shape, surface area, and applications, an SLM can be further classified as flat sheet, hollow fibre, and spiral wound SLM. Flat sheet SLM (FSSLM) is the simplest form, uses a sheet to incorporate the solvent, and can be placed between two half compartments containing the feed and receiving solutions which are under continuous stirring. HFSLM consists of a hollow fibre module made of a single non-porous material. Inside this, many thin fibres are packed in rows through which the feed solution passes and the receiving phase outside. The solid support in SLM must be hydrophobic in nature to be able to retain the organic solvent in the membrane pores by capillary action and should have excellent thermal chemical stability. Polymers such as polytetrafluoroethene (PTFE), polypropylene, and polysulphones are generally used for this purpose. The extractant used for SLM is an organic solvent, which has specificity for components and can cause extraction by complex formation, ion pair formation, or solvation. The extractants used for extraction by complex formation may be chelating, such as LIX 84-I, LIX 64N, LIX 62N, and LIX 860, or acidic, such as D2EHPA, PC88A, and Cyanex-272 (phosphoric acid derivatives). Extraction by ion pair formation is common for amine-based extractants, which extract acid and then by anion exchange reaction extract metal ions, e.g. Alamine 336, Aliquat 336, and Alamine 304. Solvating extractants are weakly basic in nature, and thus, they extract either neutral metal complexes or acids by forming a solvate. A very good example is crown ethers, which are cyclic polyethers, but as they are costly, they are rarely used. The other extractants that are commonly used are tri-n-octylphosphine oxide (TOPO), TBP, and MIBK. The diluents are generally used for the preparation of various concentrations of organic extractants, and they should have dielectric constant, low viscosity, should be cheap, etc. Diluents like xylene, toluene, hexane, and cyclohexane are generally used. The different steps involved are diffusion of metal ions from the bulk of the feed phase to the inner surface of the membrane, proton diffusion from the inner surface to the bulk of the feed phase, metal complex formation, and its diffusion from the inner to the outer surface of the membrane, wherein when in contact with the receiving phase, metal ions are stripped out while protons combine with the carrier and regenerate it, which diffuses back into the inner surface of the membrane. The stripped metal ion diffuses to the bulk of the receiving phase [15].

ELM was invented in 1968 and consists of an emulsion (a mixture of two or more immiscible liquids) of the organic membrane and aqueous receiving phases [16]. This is then dispersed into the continuous aqueous feed phase. The mass transfer of the solutes towards the internal receiving phase occurs, and it is simpler than in the conventional process, energy efficient, and compact in size. ELMs are powerful and have a large number of applications. The main limitation of ELMs is the stability of the emulsion. ELMs have many advantages like high efficiency, large interfacial area, continuous simple operation, and simultaneous extraction and stripping.

3.3 MECHANISM OF SEPARATION

Separation through membranes can proceed via either chemical or physical processes. Separations that are based on chemical processes involve chemical interactions between the solute of interest and the membrane components. This is quite common in separations using ion exchange solid membranes or liquid membranes and biological membranes. Since the latter is out of scope of the present discussion, our focus will remain on liquid membranes and solid membranes. Separation with membranes can be either equilibrium or non-equilibrium processes, which are further controlled or not controlled by pressure [17]. Separations based on liquid membranes and pressure-driven membrane distillation are equilibrium processes, whereas pressure-driven membrane filtration and non-pressure-driven electrodialysis are examples of non-equilibrium processes. A schematic representation of these processes is presented in Figure 3.4.

3.3.1 MECHANISM IN SOLID MEMBRANES

Separation processes can be driven by pressure (MF, UF, NF, and RO), diffusion (dialysis), vacuum (membrane distillation), and electrical gradient (electrodialysis). Separation in membranes can be achieved by sieving (based on sizes of solute and membrane pores) or diffusion through membrane [18]. Two basic models can be used to understand the mass transfer through membranes, namely, the solution-diffusion model and the hydrodynamic model. In actual studies, it is possible that both of these mechanisms can operate. According to the solution-diffusion model, the transport of solutes from the feed to the receiving phase occurs only by diffusion across the membrane phase. It is also assumed that the two phases are in equilibrium with the adjacent membrane phase. This principle is more important for dense membranes without natural pores. During the filtration

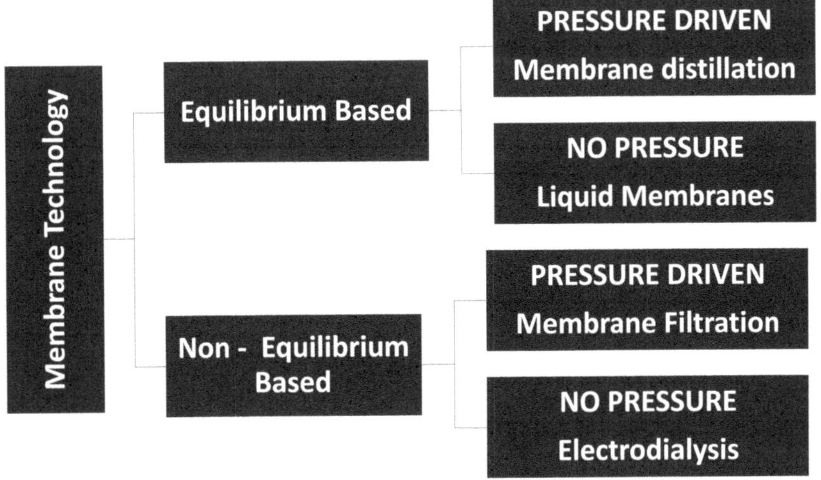

FIGURE 3.4 Schematic representation of membrane-based separation

process, a boundary layer forms on the membrane. This concentration gradient is created by molecules which cannot pass through the membrane. The effect is referred to as concentration polarization and, occurring during the filtration, leads to a reduced transmembrane flow (flux). Concentration polarization is, in principle, reversible by cleaning the membrane, which results in the initial flux being almost totally restored. Using a tangential flow to the membrane (cross-flow filtration) can also minimize concentration polarization. According to the hydrodynamic model, transport through pores, in the simplest case, is through convection. This requires the size of the pores to be smaller than the diameter of the two separate components. Membranes which function according to this principle are mainly used in microfiltration and UF. They are used to separate macromolecules from solutions and colloids from a dispersion, or remove bacteria. During this process, the retained particles or molecules form a pulpy mass (filter cake) on the membrane, and this blockage of the membrane hampers the filtration. This blockage can be reduced by the use of the cross-flow method (cross-flow filtration). Here, the liquid to be filtered flows along the front of the membrane and is separated by the pressure difference between the front and back of the membrane into the retentate (the flowing concentrate) on the front and the permeate (filtrate) on the back. The tangential flow on the front creates a shear stress that cracks the filter cake and reduces fouling. In electrically charged membranes, permselectivity is the main property that governs separation. The term permselectivity refers to the selective permeability of solutes of oppositely charged ions. A polymeric matrix with covalently bound ionizable groups swells with water when immersed. The charges are dissociated, and the counter ions can be easily exchanged with ions of interest. Thus, permeation can occur through the membranes. The flux can be affected by different conditions [19]. Figure 3.5 gives a general idea of separations using neutral and cation exchange membranes.

3.3.2 Mechanism in liquid membranes

The transport of solutes through liquid membranes is classified based on the role of the membrane phase. The membrane may be a simple solvent for the solute or for the carrier, which will complex with metal ions of interest [20]. The membrane liquid solubilizes the permeate from the feed and behaves like a solvent. The removal of the permeate from the feed is dependent on the solubility of the permeate in the membrane liquid. The transport is governed by the concentration gradient and occurs till the concentrations in both the feed and receiving phases are equal. This is known as a simple transfer process. However, if the receiving phase contains a reagent that combines with the solute of interest, then there is a continuous transport against the concentration gradient, and this is known as "uphill transport". The presence of a carrier in the membrane liquid results in the formation of a metal–carrier complex, which is then transported through the membrane phase. This is known as carrier-facilitated transport. When a permeate is transported using a carrier and, during its transport, if an equivalent amount of another species is co-transported from the receiving to the feed phase, the process is known as carrier-facilitated coupled transfer.

The metal ion transport can be represented by two different models. The most recurrently used model [21] deals with concentration diffusion layers. In this, the

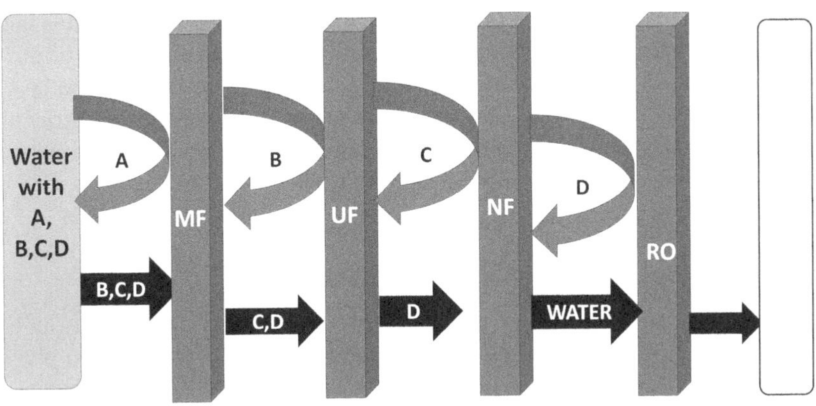

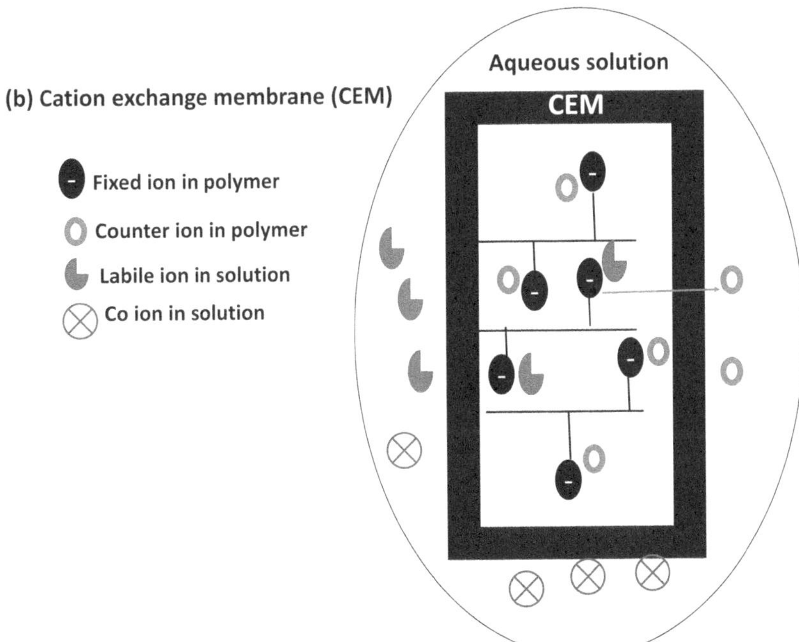

FIGURE 3.5 Schematic representation of separation using a solid membrane based on (a) sieving mechanism and (b) cation exchange

aqueous–organic interphase is treated as a platform for the processes of complexation and decomplexation to occur. The transport can be divided into various steps. The solute traverses the diffusion boundary layer (DBL) from the bulk of the feed to the membrane surface with a linear concentration gradient. The metal–carrier complexation at the feed/membrane interphase is faster than the diffusion of ions. The complex formed disperses through the organic layer by convection due to continuous stirring and reaches the membrane/receiving interphase, where the complex breaks and the metal ion again has to traverse through the DBL with a linear concentration change. The transport of solutes occurs along with the simultaneous co-transfer of different ions of the same charge from the receiving to the feed phase. The second approach that is less accepted is the "Big Carrousel" mechanism [22], wherein the carrier moves slightly out from the organic membrane in the aqueous reaction layer and then transfers from one aqueous phase to another through the membrane before finally moving back.

3.4 APPLICATIONS

Solid membranes are used in various processes such as reverse osmosis, microfiltration, and UF, wherein separation is based on membrane pore size, or in processes such as electrodialysis and fuel cells, where electrically charged membranes are used. Membrane filtration process is dependent on the size of the pores and solute to be separated. Adsorptive membranes operate on the basis of electrical double layer theory [23]. The separation efficiency of UF membranes obtained in its pristine form [24] can be enhanced using polymers [25, 26], micelles [27, 28], or powder [29, 30] as an alternative membrane technology to obtain safe drinking water. The enhanced technologies tend to reduce the formation of toxic species and also reduce leakage. Adsorptive membranes prove to be a benign alternative and reduce the risk of leakage and production of toxic species. This process is known as "membrane adsorptive filtration" [31] (or adsorptive membrane filtration [32]) and offers advantages of high removal efficiency with a high permeability flux and low operating pressure. The technique allows easy regeneration and is compact [33]. Adsorptive membranes or membrane sorbents [34, 35] tend to behave as affinity membranes [36], which tend to bind to metal ions by chelation [37, 38] or ion exchange [39]. To enhance selectivity, ion-imprinted membranes (with a specific ion template) [40] or mixed matrix membranes (MMMs; contain polymer or inorganic fillers or grains in the membrane matrix) [41, 42] are used. These membranes contain pores in addition to the functional groups [43, 44] and can be used for membrane filtration processes. Generally, polymeric filtration membranes show low affinity towards metal ions, and they need to be modified. Polyvinylidene difluoride (PVDF) membranes incorporated with amine-functionalized MCM-41 could be used for the removal of Cu(II) at near-neutral pH [45]. Novel polyethersulphone (PES)/hydrous manganese dioxide (HMO) UF MMMs were found to be useful for the removal of Pb(II) [46]. There are large numbers of studies reported on the use of similar modified membranes for the removal of metal ions [47–88]. It is seen that different materials like g-PAAm(polyacrylamide); iminodiacetate; hyper-branched poly(amido amine) 8-hydroxyquinoline; peracetic acid (PAA); polybenzimidazole; modified proclavaminic acid (PCV); and oxides of Al, Ti, Fe, etc., could be used as sorbents into membranes made of polyethylene (PE), PTFE, cellulose acetate (CA), PVDF, etc. It is seen that in most of the studies, the separation could be achieved under mild pH conditions, and elution of the separated ions was also achieved using dilute acids or complexing agents. The membranes can be classified as mixed matrix adsorptive membranes, pore-filled adsorptive membranes, and surface adsorptive membranes.

Ion exchange membranes are most commonly used for the treatment of brackish water. Membrane electrofiltration is the amalgamation of membrane and filtration technologies. The problem of membrane fouling is a challenge in this process [89]. The removal of heavy metal ions is an important application of electrodialysis [90–92]. The transport property of the membrane and the effect of various conditions were analysed on the removal of metal ions such as Cu and Zn [93–116].

Nafion ionomer membrane is an excellent ionomer with a large number of applications, e.g. as an electrochemical separator and catalysts. Extensive studies have been carried out on the structural elucidation of the membrane and the transport properties in line with its industrial applications. Nonetheless, it has not been used to a great extent for the separation of toxic cationic species. The selective permeation of $Cu^{2+\circ}$ and $UO_2^{2+\circ}$ in the presence of common cations through a Nafion membrane [117] was studied in detail using appropriate masking agents [117]. It was found that the masking agents could result in either a positive or negative effect on selectivity. Permeation of some nitrogen heterocyclic bases could be achieved using metal ion-loaded Nafion membranes, and it was observed that the presence of certain ions showed an enhancement of permeation flux [118]. The studies showed the effect of metal ions and also the structure of solutes on the permeation rate. The separation of some binary mixtures under favourable conditions was achieved. Permselectivity of Nafion membranes could be altered by structural modification, and permeation of common anions could be achieved [119]. The nature of modification and the size of anions were found to have a strong influence on permeation rates. Ion exchange studies of alkali and transition metal ions were carried out using Nafion [117] membranes in the protonic form [120, 121]. It was observed that the water sorption isotherms obtained were quite different. The results showed an alteration of the physical structure of the exchangers upon long-storage or ageing. It was also observed that the amount of water sorbed was low and the counterions were less hydrated, resulting in high discrepancy in the ion exchange behaviour. Pretreatment of Nafion [117] membranes resulted in the modification of the membrane structure and arrangement of ionogenic groups. Thus, the state of water present in the membrane was quite different, and it altered the ion exchange property of the membrane. It can be concluded that the ageing of ion exchangers results in the loss of their elasticity and showed poor water uptake. To get a better understanding, the values of swelling pressure and free energy of swelling were calculated. It was observed that the presence of transition metal ions in Nafion resulted in a higher value of free energy of swelling compared to the protonated form. This could be correlated to increased hydration of metal ion-incorporated membranes. The uptake of metal ions was subjected to modelling using different isotherm models, and it was established that the process was chemisorption. The role of pretreatment was also found to be predominant. The permeation of transition metal ions $\left(Cu^{2+}, Co^{2+}, Ni^{2+}, Zn^{2+}\right)$ through H^+ and alkali metal ion forms of Nafion membrane was carried out. It was observed that the diffusion coefficient values (D) were directly proportional to the individual ion exchange selectivity values (K). Thus, it could be established that the initial stage of permeation is governed by the ion exchange process. The uptake of transition metal ions $\left(Cu^{2+}, Co^{2+}\right)$ by Nafion membrane was carried out in the presence of different complexing organic reagents or ligands. This was carried out to understand how uptake can be affected in a sample mixture. The studies were also carried out using differently pretreated Nafion membranes. It was observed that the strength and the charge of the metal–ligand complexes had an effect on the uptake capacity [122]. In order to get an insight of the uptake and permeation process, the studies were carried out using two cationic dyes [123]. In this study, the Nafion membrane was pretreated in two different ways prior to its use. Different characterization techniques were used to understand the changes. The presence of water clusters could be identified by spectrophotometric and thermal measurements, and the changes in the water cluster upon pretreatment were also established. It was also possible to clearly see the differences in the inter-actions of the two dyes with the membranes. Image analysis of the dye-loaded mem-branes could clearly depict the structural changes due to pretreatment. Permeation studies established the dependence of permeability and diffusion coefficients on the nature of the dye and the pretreatment protocols adopted [123].

Membrane pertraction methods can have applications at the lab scale and at the industrial level. BLMs can be used in the laboratory to test the carrier activity of various compounds (well-known gravimetric reagents or novel complexing agents). The reports in the literature indicate that a massive amount of studies have been car-ried out using BLMs for the transport of metals, ions, organic dyes, etc. Therefore, it is pragmatic to restrict the discussion to few examples to give an overall idea to the readers. The use of 8-hydroxyquinoline (oxine) as a carrier has demonstrated the pos-sibility of achieving selective metal ion transport using a well-known non-selective gravimetric reagent as the carrier [124,125]. The main approach is to modify the solution conditions. The use of amino acids or other reagents results in enhanced metal ion transport due to synergistic effects. BLM studies using newly synthesized phenoxy ethers demonstrated a high degree of selectivity [126]. The mechanism of transport was carrier-facilitated coupled transport, wherein protons migrated in the direction opposite to the alkali metal ions. The transport of radionuclides like uranium using BLMs could be achieved using carriers like potassium ion selective crown ethers [127] and calixarenes [128]. When crown ether was used as a car-rier, the uranyl ion was converted to its anionic uranyl thiocyanate complex using potassium thiocyanate (KSCN) at a pH of 1 and thus transported along with the potassium ions. Masking agents could be used to remove interferences from dif-ferent metal ions. The main limitation was the transport was achieved only in the presence of KSCN at a pH of 1. It was observed that the limitations of low pH and interferences could be eradicated using uranyl ion-selective calixarenes as carriers. The studies showed that in the presence of TOPO as synergistic reagent, more than 95% transport could be achieved, and this was applied to actual seawater samples [128]. The transport of uranium using a mixture of thenoyltrifluoroacetone (HTTA) and dicyclohexyl-18C6 as carriers was nearly complete without any interferences from most of the ions except Th(IV) and Cu^{2+}. The use of masking agents further enhanced the selectivity of separation [129].

SLMs have received considerable interest due to their ease of operation, high selectivity, and low operating cost [130]. They have been used extensively for the separation and pre-concentration of metal ions [131–160]. It has been observed that different solvents like kerosene, toluene, xylene, and dodecane can be used as sol-vents like chloroform. Several research studies are being carried out at the laboratory and pilot plant levels for the use of ELMs for the removal of metal ions [161,162]. Membrane processes can be used for the removal of dyes [163,164]. The selective removal of malachite green dye from waste water using SLMs has been reported [165]. The separation of reactive dyes, namely, Gold Yellow (GYHE-R) and Reactive Green HE4BD (RGHE-4BD), has been achieved using Aliquot-336 carrier [166]. The extraction of Congo red (CR), an anionic disazo direct dye, from aqueous solutions by using ELMs has been carried out [167]. Salicylic acid in benzene has been used for the removal of methylene blue dye from textile waste water [168], while di-(2-ethyl hexyl) phosphoric acid (D2EHPA) as carrier could be used for the transport of methyl violet and rhodamine B [169]. The anionic dye Cibacron Red FN-R was separated using a BLM using tetrabutylammonium bromide (TBAB) carrier dissolved in methylene chloride solvent [170].

3.5 CONCLUSIONS

Membrane processes are non-destructive methods. Membrane filtration methods have matured to a great extent, with a wide range of membranes available for applications. This technique is simple, rapid, and efficient even at high concentrations. The quality of effluent treated is quite high. The main disadvantages are the low throughput with a limited flow rate and is not suited for low concentration levels. Membrane selection is based on the application, making this process more cumbersome. In addition to this, the high installation and maintenance costs coupled with high energy requirements make this process quite unwieldly. Solid membranes are also prone to fouling, leading to deterioration of membrane efficiency.

To overcome these difficulties and to have an effective separation method, especially for low concentration levels, sorption seems to be quite an attractive method. This method is discussed in the next chapter.

REFERENCES

1. J. Glater, *Desal.* 117 (1998) 297–309.
2. A. Nollet, *Lecons de Physique Experimentale*, Chez les Frères Guérin, 1748.
3. E.M. Hoek, V.V. Tarabara, and J. Kucera, Membrane Materials and Module Development, Historical Perspective, In *Encyclopedia of Membrane Science and Technology*, Ed. E.M. Hoek and V.V. Tarabara, John Wiley & Sons, Inc., 2013, https://doi.org/10.1002/9781118522318.emst033.
4. IUPAC, *Compendium of Chemical Terminology*, 2nd ed. (The "Gold Book"), Compiled A.D. McNaught and A. Wilkinson, Blackwell Scientific Publications, 1997, Online version (2019-) created by S. J. Chalk, ISBN 0-9678550-9-8, https://doi.org/10.1351/goldbook.
5. P.K. Parhi, *J. Chem.* 2013 (2013) Article ID 618236, 11 pages, https://doi.org/10.1155/2013/618236.
6. R.H. Perry and D.H. Green, *Perry's Chemical Engineers' Handbook*, 7th ed., McGraw-Hill, 1997.
7. M. De Falco, A. Salladini, E. Palo, and G. Iaquaniello, Reformer and Membrane Modules (RMM) for Methane Conversion Powered by a Nuclear Reactor, In *Nuclear Power – Deployment, Operation and Sustainability*, Ed. P. Tsvetkov, September 2011 InTech, https://doi.org/10.5772/20603.

8. Jayshree Ramkumar, Nafion Perfluorosulphonate Membrane: Unique Properties and Various Applications, In *Functional Materials*, Ed. S. Banerjee and A.K. Tyagi, Elsevier Inc., 2012.

9. C. Bhattacharjee, V. Saxena, and S. Dutta, *Innov. Food Sci. Emerg. Technol.* 43 (2017), https://doi.org/10.1016/j.ifset.2017.08.002.

10. S. Björkegren, R.F. Karimi, A. Martinelli, N.S. Jayakumar, and M.A. Hashim, *Membr.* 5 (2015) 168–179, https://doi.org/10.3390/membranes5020168.

11. Jayshree Ramkumar, and S. Chandramouleeswaran, *Ind. J. Adv. Chem. Sci.* 3 (2015) 293–298.

12. V.S. Kislik, Bulk Liquid Membrane, In *Encyclopedia of Membranes*, Ed. E. Drioli and L. Giorno, Springer, 2012, https://doi.org/10.1007/978-3-642-40872-4_83-1.

13. S. Chang, *Des Wat. Treat.* (2015) 1–9, https://doi.org/10.1080/19443994.2015.1102772.

14. V.S. Kislik, Introduction, Chapter 1, In *Liquid Membranes: Principles and Applications in Chemical Separations and Wastewater Treatment*, Ed. V.S. Kislik, Elsevier Science, 2009.

15. V.S. Kislik, Supported Liquid Membrane, In *Encyclopedia of Membranes*, Ed. E. Drioli and L. Giorno, Springer, 2014, https://doi.org/10.1007/978-3-642-40872-4_566-2.

16. A. Kumar, A. Thakur, and P.S. Panesar, *Rev. Environ. Sci. Biotechnol.* 18 (2019) 153–182, https://doi.org/10.1007/s11157-019-09492-2.

17. E. OboteyEzugbe and S. Rathilal, *Membr.* 10 (2020) 89, https://doi.org/10.3390/membranes10050089.

18. T.A. Saleh and V.K. Gupta, Chapter 1 – An Overview of Membrane Science and Technology, In *Nanomaterial and Polymer Membranes*, Ed. T.A. Saleh and V.K. Gupta, pp. 1–23, Elsevier, 2016.

19. P. Meares, Transport in Ion Exchange Membranes, In *Synthetic Membranes: Science, Engineering and Applications*, Ed. P.M. Bungay, H.K. Lonsdale, and M.N. de Pinho, NATO ASI Series (Series C: Mathematical and Physical Sciences), Vol. 181, Springer, 1986.

20. V.S. Kislik, Transport Mechanisms with Liquid Membranes, In *Encyclopedia of Membranes*, Ed. E. Drioli and L. Giorno, Springer, 2016, https://doi.org/10.1007/978-3-662-44324-8_588.

21. C.F. Reusch and E.L. Cussler, *AIChE J.* 19 (1973) 736–741, https://doi.org/10.1002/aic.690190409.

22. V. Mogutov and N.M. Kocheriginsky, *J. Membr. Sci.* 79 (1993) 273–283.

23. Z.-Q. Huang and Z.-F. Cheng, *J. Appl. Polym. Sci.* 137 (2020) 48579, https://doi.org/10.1002/app.48579.

24. J.M. Arnal, B. Garcia-Fayos, G. Verdu, and J. Lora, *Desal.* 248 (2009) 34–41.

25. R. Kumar, A.M. Isloor, and A. Ismail, *Desal.* 350 (2014) 102–108.

26. P. Kanagaraj, A. Nagendran, D. Rana, T. Matsuura, S. Neelakandan, and T. Karthikkumar, *Appl. Surf. Sci.* 329 (2015) 165–173.

27. X. Zhang, L. Niu, F. Li, X. Zhao, and H. Hu, *Sep. Purif. Technol.* 175 (2017) 314–320.

28. P. Bahmani, A. Maleki, H. Daraei, M. Khamforoush, R. Rezaee, F. Gharibi, A.G. Tkachev, A.E. Burakov, S. Agarwal, and V.K.J. Gupta, *J. Colloid Interf. Sci.* 506 (2017) 564–571.

29. N. Yin, K. Wang, L. Wang, and Z. Li, *Chem. Eng. J.* 306 (2016) 619–628.

30. L. Hao, T. Zheng, J. Jiang, G. Zhang, and P. Wang, *Chem. Eng. J.* 292 (2016) 163–173.

31. Y.P. Tang, L. Luo, Z. Thong, and T.S. Chung, *J. Membr. Sci.* 541 (2017) 434–446.

32. S. Ali, S.A.U. Rehman, I.A. Shah, M.U. Farid, A.K. An, and H. Huang, *J. Hazard. Mater.* 365 (2019) 64.

33. Y. Bao, X. Yan, W. Du, X. Xie, Z. Pan, J. Zhou, and L. Li, *Chem. Eng. J.* 281 (2015) 460–467.

34. D. Bhattacharyya, J.A. Hestekin, S.M.C. Ritchie, and L.G. Bachas, *Membr. Technol.* 110 (1999) 8–11.

35. E. Salehi, P. Daraei, and A.A. Shamsabadi, *Carbohydr. Polym.* 152 (2016) 419–432.

36. E. Klein, *J. Membr. Sci.* 179 (2000) 1–27.

37. L. Lebrun, F. Vallée, B. Alexandre, and Q.T. Nguyen, *Desal.* 207 (2007) 9–23.

38. J. Meng, J. Yuan, Y. Kang, Y. Zhang, and Q. Du, *J. Coll. Interf. Sci.* 368 (2012) 197–207.

39. M.M. Nasef and A.H. Yahaya, *Desal.* 249 (2009) 677–681.

40. E. Salehi, S.S. Madaeni, and V. Vatanpour, *J. Membr. Sci.* 389 (2012) 334–342.

41. R.J. Gohari, W.J. Lau, T. Matsuura, E. Halakoo, and A.F. Ismail, *Sep. Purif. Technol.* 120 (2013) 59–68.

42. A.R. Kamble, C.M. Patel, Z.V.P. Murthy, *Ren. Sustain. Ener. Rev.* 145 (2021) 111062.

43. A. Ghaee, M. Shariaty-Niassar, J. Barzin, and T. Matsuura, *Chem. Eng. J.* 165 (2010) 46–55.

44. A. Ghaee, M. Shariaty-Niassar, J. Barzin, and A. Zarghan, *App. Surf. Sci.* 258 (2012) 7732–7743.

45. Y. Bao, X. Yan, W. Du, X. Xie, Z. Pan, J. Zhou, and L. Li, *Chem. Eng. J.* 281 (2015) 460–467.

46. R.J. Gohari, W.J. Lau, T. Matsuura, E. Halakoo, and A.F. Ismail, *Sep. Purif. Technol.* 118 (2013) 64–72.

47. B. Gupta, N. Anjum, and K. Sen *J. Appl. Polym. Sci.* 85 (2002) 282–291.

48. S. Tsuneda, K. Saito, and S. Furusaki, *J. Membr. Sci.* 58 (1991) 221–234.

49. H. Yoo and S.Y. Kwak, *J. Membr. Sci.* 448 (2013) 125–134.

50. A. Denizli, D. Tanyolac, B. Salih, E. Aydinlar, A. Ozdural, and E. Piskin, *J. Membr. Sci.* 137 (1997) 1–8.

51. D. Bhattacharyya, J.A. Hestekin, P. Brushaber, L. Cullen, L.G. Bachas, and S.K. Sikdar, *J. Membr. Sci.* 141 (1998) 121–135.

52. E. Salehi, S.S. Madaeni, and F. Heidary, *Sep. Purif. Technol.* 94 (2012) 1–8.

53. M. Kumar, R. Shevate, R. Hilke, and K.V. Peinemann, *Chem. Eng. J.* 301 (2016) 306–314.

54. E. Salehi, S.S. Madaeni, L. Rajabi, A.A. Derakhshan, S. Daraei, and V. Vatanpour, *Chem. Eng. J.* 215–216 (2013) 791–801.

55. L. Zhang, Y.H. Zhao, and R. Bai, *J. Membr. Sci.* 379 (2011) 69–79.

56. H. Bessbousse, J.F. Verchère, and L. Lebrun, *J. Membr. Sci.* 364 (2010) 167–176.

57. C.O. M'Bareck, Q.T. Nguyen, S. Alexandre, and I. Zimmerlin, *J. Membr. Sci.* 278 (2006) 10–18.

58. W.P. Zhu, S.P. Sun, J. Gao, F.J. Fu, and T.S. Chung, *J. Membr. Sci.* 456 (2014) 117–127.

59. V. Nayak, M.S. Jyothi, R.G. Balakrishna, M. Padaki, and A.M. Isloor, *RSC Adv.* 6 (2016) 25492–25502.

60. V. Nayak, M.S. Jyothi, R.G. Balakrishna, M. Padaki, and S. Deon, *J. Hazard. Mater.* 331 (2017) 289–299.

61. K.N. Han, B.Y. Yu, and S.Y. Kwak, *J. Membr. Sci.* 396 (2012) 83–91.

62. M.R. Kotte, A.T. Kuvarega, M. Cho, B.B. Mamba, and M.S. Diallo, *Environ. Sci. Technol.* 49 (2016) 9431–9442.

63. C. Magnenet, S. Lakard, C.C. Buron, and B.J. Lakard, *J. Colloid Interface Sci.* 376 (2012) 202–208.

64. S. Lakard, C. Magnenet, M.A. Mokhter, M. Euvrard, C.C. Buron, and B. Lakard, *Sep. Purif. Technol.* 149 (2015) 1–8.

65. N. Ghaemi, *Appl. Surf. Sci.* 364 (2016) 221–228.

66. X. Zhang, Y. Wang, Y. You, H. Meng, J. Zhang, and X. Xu, *Appl. Surf. Sci.* 263 (2012) 660–665.

67. M. Delavar, G. Bakeri, and M. Hosseini, *Chem. Eng. Res. Des.* 120 (2017) 240–253.

68. N. Abdullah, R.J. Gohari, N. Yusof, A.F. Ismail, J. Juhana, W.J. Lau, and T. Matsuura, *Chem. Eng. J.* 289 (2016) 28–37.

69. J. Liu, L. Zheng, Y. Li, M. Free, and M.Z. Yang, *RSC Adv.* 6 (2016) 51757–51767.

70. Y. Yurekli, *J. Hazard. Mater.* 309 (2016) 53–64.

71. Y. Yurekli, M. Yildirimb, L. Aydinc, and M. Savran, *J. Hazard. Mater.* 332 (2017) 33–41.
72. S. Pan, J. Li, O. Noonan, X. Fang, G. Wan, C. Yu, and L. Wang, *Environ. Sci. Technol.* 51 (2017) 5098–5107.
73. M. Mondal, M. Dutta, and S. De, *Sep. Purif. Technol.* 188 (2017) 155–166.
74. K.H. Chan, E.T. Wong, A. Idris, and N.M. Yusof, *J. Ind. Eng. Chem.* 27 (2015) 283–290.
75. K.H. Chan, E.T. Wong, M. Irfan, A. Idris, and N.M. Yusof, *J. Taiwan Inst. Chem. E* 47 (2015) 50–58.
76. S.S. Madaeni, S. Zinadini, and V. Vatanpour, *Sep. Purif. Technol.* 80 (2011) 155–162.
77. M.U. Farid, H.-Y. Luan, Y. Wang, H. Huang, A. Kyoungjin, and R. Jalil Khan, *Chem. Eng. J.* 325 (2017) 239–248.
78. P. Tan, J. Sun, Y. Hu, Z. Fang, Q. Bi, Y. Chen, and J. Cheng, *J. Hazard. Mater.* 297 (2015) 251–260.
79. R. Mukherjee, P. Bhunia, and S. De, *Chem. Eng. J.* 292 (2016) 284–297.
80. N. Ghaemi, S. Zereshki, and S. Heidari, *Process Saf. Environ.* 111 (2017) 475–490.
81. G. Zeng, Y. He, Y. Zhan, L. Zhang, Y. Pan, C. Zhang, and Z. Yu, *J. Hazard. Mater.* 317 (2016) 60–72.
82. N. Ghaemi, S.S. Madaeni, P. Daraei, H. Rajabi, A. Alizadeh, R. Heydari, M. Beygzadeh, S. Zinadini, and S. Ghouzivand, *Chem. Eng. J.* 263 (2015) 101–112.
83. N. Ghaemi, and P. Daraei, *J. Ind. Eng. Chem.* 40 (2016) 26–33.
84. J. Song, O. Hyuntaek, H. Kong, and J. Jang, *J. Hazard. Mater.* 187 (2011) 311–317.
85. M. Zhu, L. Zhu, J. Wang, T. Yue, R. Li, and Z. Li, *J. Environ. Manag.* 196 (2017) 127–136.
86. N. Saffaj, H. Loukili, S.A. Younssi, A. Albizane, M. Bouhria, M. Persin, and A. Larbot, *Desal.* 168 (2004) 301–306.
87. V.V. Tomina, N.V. Stolyarchuk, I.V. Melnyk, and V.M. Kochkodan, *Micropor. Mesopor. Mater.* 209 (2015) 66–71.
88. V. Namboodiri and N. Rajagopalan, 2.6 – Desalination, In *Comprehensive Water Quality and Purification*, Ed. S. Ahuja, pp. 98–119, Elsevier, 2014, https://doi.org/10.1016/B978-0-12-382182-9.00095-5.
89. M.A. Barakat, *Arab. J. Chem.* 4 (2011) 361–377, https://doi.org/10.1016/j.arabjc.2010.07.019.
90. T. Scarazzato, Z. Panossian, J.A.S. Tenório, V. Pérez-Herranz, and D.C.R. Espinosa, *J. Clean. Prod.* 168 (2017) 1590–1602, https://doi.org/10.1016/j.jclepro.2017.03.152.
91. Ö. Arar, Ü. Yüksel, N. Kabay, and M. Yüksel, *Desal.* 342 (2014) 16–22, https://doi.org/10.1016/j.desal.2014.01.028.
92. T. Benvenuti, M. García-Gabaldón, E.M. Ortega, M.A.S. Rodrigues, A.M. Bernardes, V. Pérez-Herranz, and J. Zoppas-Ferreira, *J. Membr. Sci.* 542 (2017) 320–328.
93. M. Martí-Calatayud, M. García-Gabaldón, and V. Pérez-Herranz, *Currents. Appl. Sci.* 8 (2018) 1566.
94. N. Nemati, S.M. Hosseini, and M. Shabanian, *J. Hazard. Mater.* 337 (2017) 90–104.
95. M.C. Martí-Calatayud, M. García-Gabaldón, and V. Pérez-Herranz, *J. Membr. Sci.* 443 (2013) 181–192.
96. P. Sharma and V.K. Shahi, *Chem. Eng. J.* 382 (2020) 122688.
97. C. Vallois, P. Sistat, S. Roualdès, and G. Pourcelly, *J. Membr. Sci.* 216 (2003) 13–25.
98. J.H. Chang, C.P. Huang, S.F. Cheng, and S.Y. Shen, *Process. Saf. Environ. Prot.* 112 (2017) 235–242.
99. K.S. Barros, T. Scarazzato, and D.C.R. Espinosa, *Sep. Purif. Technol.* 193 (2018) 184–192.
100. A. Mahmoud, L. Muhr, S. Vasiluk, A. Aleynikoff, and F. Lapicque, *J. Appl. Electrochem.* 33 (2003) 875–884.
101. T. Scarazzato, Z. Panossian, M. García-Gabaldón, E.M. Ortega, J.A.S. Tenório, V. Pérez-Herranz, and D.C.R. Espinosa, *J. Membr. Sci.* 535 (2017) 268–278.

102. F. Aouad, A. Lindheimer, and C. Gavach, *J. Membr. Sci.* 123 (1997) 207–223.

103. M.A.S. Rodrigues, F.D.R. Amado, M.R. Bischoff, C.A. Ferreira, A.M. Bernardes, and J.Z. Ferreira, *Desal.* 227 (2008) 241–252.

104. R.F. Dalla Costa, M. AntônioSiqueira Rodrigues, and J.Z. Ferreira, *Sep. Sci. Technol.* 33 (1998) 1135–1143.

105. M.E. Vallejo, F. Persin, C. Innocent, P. Sistat, and G. Pourcelly, *Sep. Purif. Technol.* 21 (2000) 61–69.

106. M.A.S. Rodrigues, R.F. Dalla Costa, A.M. Bernardes, and J. Zoppas Ferreira, *Electrochim. Acta* 47 (2001) 753–758.

107. Y. Çengeloğlu, A. Tor, E. Kir, and M. Ersöz, *Desal.* 154 (2003) 239–246.

108. S.M. Hosseini, S. Sohrabnejad, G. Nabiyouni, E. Jashni, B. Van der Bruggen, and A. Ahmadi, *J. Membr. Sci.* 583 (2019) 292–300.

109. E. Jashni and S.M. Hosseini, *Sep. Purif. Technol.* 234 (2020) 116118.

110. S.M. Hosseini, H. Alibakhshi, E. Jashni, F. Parvizian, J.N. Shen, M. Taheri, M. Ebrahimi, and N. Rafiei, *J. Hazard. Mater.* 381 (2020) 120884.

111. K.S. Barros and D.C.R. Espinosa, *Sep. Purif. Technol.* 201 (2018) 244–255.

112. T. Mohammadi, A. Moheb, M. Sadrzadeh, and A. Razmi, *Sep. Purif. Technol.* 41 (2005) 73–82.

113. M. Sadrzadeh, A. Razmi, and T. Mohammadi, *Sep. Purif. Technol.* 54 (2007) 147–156.

114. S. Itoi, I. Nakamura, and T. Kawahara, *Desal.* 32 (1980) 383–389.

115. T. Benvenuti, R.S. Krapf, M.A.S. Rodrigues, A.M. Bernardes, and J. Zoppas-Ferreira, *Sep. Purif. Technol.* 129 (2014) 106–112.

116. Jayshree Ramkumar, K.S. Shrimal, B. Maiti, and T.S. Krishnamoorthy, *J. Memb. Sci.* 116 (1996) 31–37.

117. Jayshree Ramkumar, B. Maiti, and T.S. Krishnamoorthy, *J. Memb. Sci.* 125 (1997) 269–274.

118. Jayshree Ramkumar, E.K. Unnikrishnan, B. Maiti, and P.K. Mathur, *J. Memb. Sci.* 141 (1998) 283–288.

119. Jayshree Ramkumar and T. Mukherjee, *Sep. Purif. Technol.* 54 (2007) 61–70.

120. Jayshree Ramkumar and T. Mukherjee, *Talanta* 71 (2007) 1054–1060.

121. Jayshree Ramkumar and S. Chandramouleeswaran, *MOJ. Bioorg. Org. Chem.* 1 (2017) 00042, https://doi.org/10.15406/mojboc.2017.01.00042.

122. Jayshree Ramkumar, S. Chandramouleeswaran, M. Basu, and K.V.V. Devi, *J. Memb. Sci. Res.* 4 (2018) 85–92, https://doi.org/10.22079/jmsr.2018.71696.1159.

123. T. Reddy, Jayshree Ramkumar, and S. Chandramouleeswaran, *J. Memb. Sci.* 351 (2010) 11–15.

124. Jayshree Ramkumar and B. Maiti, *SST* 39 (2004) 449–457.

125. Jayshree. Ramkumar, B. Maiti, S.K. Nayak, and P.K. Mathur, *Sep. Sci. Technol.* 34 (1999) 2069–2077.

126. Jayshree Ramkumar, B. Maiti, P.K. Mathur, and K. Dhole, *Sep. Sci. Technol.* 35 (2000) 2535–2541.

127. Jayshree. Ramkumar, S.K. Nayak, and B. Maiti, *J. Memb. Sci.* 196 (2002) 203–210.

128. M. Shamsipur, R. Davarkhah, and A.R. Khanchi, *Sep. Purif. Technol.* 71 (2010) 63–69.

129. M. Amini, et al., *J. Memb. Sci. Res.* 4 (2018) 121–135.

130. S. Altin, Y. Yildirim, and A. Altin, *Hydromet.* 103 (2010) 144–149.

131. M. Shamsipur, O.R. Hashemi, and V. Lippolis, *J. Membr. Sci.* 282 (2006) 322–327.

132. A. Amiri, A. Safavi, A.R. Hasaninejad, H. Shrghi, and M. Shamsipur, *J. Membr. Sci.* 325 (2008) 295–300.

133. S.U. Rehman, G. Akhtar, M.A. Chaudry, K. Ali, and N. Ullah, *J. Membr. Sci.* 389 (2012) 287–293.

134. A. Gherrou, H. Kerdjoudj, R. Molinari, and E. Drioli, *Desal.* 139 (2001) 317–325.

135. O. Arous, *J. Membr. Sci.* 241 (2004) 177–185.

136. P.K. Bhattacharyya, T. Mohapatra, T. Gadly, D.R. Raut, S.K. Ghosh, and V.K. Manchanda, *J. Hazard. Mater.* 195 (2011) 238–244.
137. S. Ansari, P. Mohapatra, D. Prabhu, and V. Manchanda, *J. Membr. Sci.* 282 (2006) 133–136.
138. R. Güell, C. Fontàs, E. Anticó, V. Salvadó, J.G. Crespo, and S. Velizarov, *Sep. Purif. Technol.* 80 (2011) 428–434.
139. J. Lv, Q. Yang, J. Jiang, and T.-S. Chung, *Chem. Eng. Sci.* 62 (2007) 6032–6039.
140. F.J. Alguacil and H. Tayibi, *Desal.* 180 (2005) 181–187.
141. N.S. Rathore, A. Leopold, A.K. Pabby, A. Fortuny, M.T. Coll, and A.M. Sastre, *Hydromet.* 96 (2009) 81–87.
142. S. Azzoug, O. Arous, and H. Kerdjoudj, *J. Environ. Chem. Eng.* 2 (2014) 154–162.
143. P.K. Parhi, N.N. Das, and K. Sarangi, *J. Hazard. Mater.* 172 (2009) 773–779.
144. S. Gu, Y. Yu, D. He, and M. Ma, *Sep. Purif. Technol.* 51 (2006) 277–284.
145. F.J. Alguacil and P. Navarro, *Hydromet.* 61 (2001) 137–142.
146. H.G. Nowier, N. El-Said, and H.F. Aly, *J. Membr. Sci.* 177 (2000) 41–47.
147. R. Mahmoodi, T. Mohammadi, and M.K. Moghadam, *Chem. Pap.* 68 (2013) 751–756.
148. S. Altin, S. Alemdar, A. Altin, and Y. Yildirim, *Sep. Sci. Technol.* 46 (2011) 754–764.
149. J. Gega, W. Walkowiak, and B. Gajda, *Sep. Purif. Technol.* 23 (2001) 551–558.
150. M. Chaudry, N. Bukhari, M. Mazhar, and W. Abbasi, *Sep. Purif. Technol.* 55 (2007) 292–299.
151. E. Rodríguez de San Miguel, X. Vital, and J. de Gyves, *J. Hazard. Mater.* 273 (2014) 253–262.
152. B. Solangi, F. Özcan, G. Arslan, M. Ersöz, *Sep. Purif. Technol.* 118 (2013) 470–478.
153. F.J. Alguacil, A.G. Coedo, and M.T. Dorado, *Hydromet.* 57 (2000) 51–56.
154. P. Venkateswaran, *J. Environ.* 19 (2007) 1446–1453.
155. M.H.H. Mahmoud, *Sep. Purif. Technol.* 84 (2012) 63–71.
156. B. Zhang and G. Gozzelino, *Coll. Surf. A Physicochem. Eng. Asp.* 215 (2003) 67–76.
157. K. Bhatluri, M.S. Manna, A.K. Ghoshal, and P. Saha, *J. Hazard. Mater.* 299 (2015) 504–512.
158. T.K. Kaya, A. Hol, A. Surucu, and H.K. Alpoguz, *Desalin. Water Treat.* 52 (2013) 3219–3225.
159. S. Panja, P.K. Mohapatra, S.C. Tripathi, and V.K. Manchanda, *Sep. Sci. Technol.* 45 (2010) 1112–1120.
160. F.J. Alguacil and M. Alonso, *Sep. Purif. Technol.* 411 (2005) 79–184.
161. A. Kumar, A. Thakur, and P.S. Panesar, *Rev. Environ. Sci. Biotechnol.* 18 (2019) 153–182.
162. A.L. Ahmad, A. Kusumastuti, C.J.C. Derek, and O. Seng, *Chem. Eng. J.* 171 (2011) 870–882.
163. C. Thamaraiselvan and M. Noel, *Crit. Rev. Environ. Sci. Technol.* 45 (2014), https://doi.org/10.1080/10643389.2014.900242.
164. A.L. Ahmad, W. Harris, S. Syafiie, and O. Seng, *Jurnal Teknologi* 36 (2002), https://doi.org/10.11113/jt.v36.581.
165. M.W. Ashraf, N. Abulibdeh, and A. Salam, *Int. J. Environ. Res. Public Health* 16(18) (2019) 3484, https://doi.org/10.3390/ijerph16183484.
166. G. Muthuraman, *Green Chem. Technol. Lett.* 1 (2015) 54–60.
167. A. Dâas and O. Hamdaoui, *J. Haz. Mat.* 178 (2010) 973–981.
168. M. Soniya and G. Muthuraman, *J. Ind. Eng. Chem.* 30 (2015) 266–273.
169. N. Hajarabeevi, I. Mohammed Bilal, D. Easwaramoorthy, and K. Palanivelu, *Desal.* 245 (2009) 19–27.
170. G. Muthuraman and M. Ibrahim, *J. Ind. Eng. Chem.* 19 (2013) 444–449.

4 Sorption
Backbone of separation science

4.1 INTRODUCTION

Sorption is part of our daily life in many ways. Some of the simple happenings denote the sorption process. Every time we buy a new item, there is always a small silica gel packet kept in the cardboard box. This is to remove the moisture away from the item by the sorption of water vapour on the exterior of silica gel particles. Pandemic and surgical situations necessitate the use of masks, which act as a filter. Another daily application of sorption is in the process of water purification by using alum or water softening using ion exchangers. The removal of colour and impurities from water also occurs through sorption, wherein impurities are deposited on the surface of a solid. A very common example is the removal of spilt ink using cotton.

There is always an ongoing research on the development of new separation techniques which are more efficient and also cost-effective. Sorption is one such technique which is simple and also cost-effective. It is a kinetically driven process based on the differences in chemical properties of solutes. Sorption has been known for ages when different materials were used for the treatment of polluted water. Finely divided and porous materials were well-known since ancient times, and adsorption was quite prevalent in old times as can be seen from the book *Glauberusconcentratus; oder, Kern der. Glauberschen*, written by the famous alchemist Johann Rudolf Glauber (1604–1670) [1]. In the book, it is mentioned that the best precipitation is performed using special sand, the origin of which I don't like to reveal. It is able to remove all salt, slime, stench and pollution from water and liquid matter. It removes the innate redness of red wine, beer etc., etc. in few hours so that all becomes clear and bright like spring-water, so that all can be drunk well and that red wine becomes white wine. Sea water becomes sweet if it runs through sand and leaves behind all its salt. When this happens naturally, why not by arts? But the actual scientific approach was in the 1900s when isotherms were measured by J.M. van Bemmelen [1].

4.2 RUDIMENTS OF SORPTION

Sorption can be defined as a phenomenon of fixation or capture of a gas or vapour (sorbate) by a substance in the condensed state (solid or liquid) called sorbent [2]. The general definition pertains to gas sorption. Sorption phenomena involve both thermophysical and thermochemical aspects [3]. Sorption is a universal name encompassing both absorption and adsorption processes [4]. Absorption is the process in which a liquid or a gas enters into the bulk of a solid (absorbent) [4], while adsorption refers to the binding of a gas or liquid on a solid surface. Adsorption can be physical

DOI: 10.1201/9781003442516-4

(physisorption) and/or chemical (chemisorption) in nature [5]. Physisorption occurs via weak van der Waals forces of attraction and so is reversible as no new compound is formed. Chemisorption involves the formation of strong chemical bonds between the sorbate and sorbent, leading to the formation of new complexes/compounds, making this process irreversible. Generally in chemisorption, new ion pairs are formed which can be broken to retrieve the sorbent in its original form. This requires change of solution conditions, thus making this process not really reversible. The enthalpy values for physisorption lies in the range of 20–40 kJ/mol, while for chemisorption it is in the range of 40–400 kJ/mol (equivalent to chemical bond formation). Physisorption is favoured at low temperatures, while higher temperatures could suit chemisorption. The process of chemisorption is monolayer, while physisorption is multilayer in nature. Sorption is now used for the removal of species known as sorbate from aqueous solutions using solids known as sorbents. Thus, sorption can be understood as a process (physical/chemical) by which one substance is attached to another in one of the three ways, namely, absorption (assimilation of a substance present in one state into another substance of a different state, e.g. liquids sorbed by solid, and gases sorbed by liquids); adsorption (physical adherence or bonding of ions and molecules onto the surface of another phase (e.g. reagents adsorbed to the surface of a solid catalyst); and ion exchange (exchange of ions between two electrolytes or between an electrolyte solution and a solid phase). Thus, sorption can be defined as a phenomenon of fixation or capture of a sorbate (gas or ions) by a substance called sorbent (solid or liquid). The reverse of sorption is desorption. Sorption is carried out in batch mode, and the schematic representation is given in Figure 4.1. The method involves the equilibration of a fixed volume of solution with a weighed amount of sorbent for a known period. The solution contains a known concentration of solute and is maintained at a particular pH. Once the time of equilibration is over, the two phases (solid/liquid) are separated (by centrifugation/filtration), and the amount of sorbate left behind in the solution is analysed using different techniques (Figure 4.1). Various experimental conditions (pH, time, sorbent amount, solute concentration, and temperature) are optimized to attain maximum sorption.

The amount of sorbate taken up by the sorbent is denoted as q_e, which is expressed as mg/g of sorbent. It is calculated using the difference in concentrations per unit volume (mg/L) before (Co) and after equilibration (Ci) per unit mass of sorbent using Eq. 1. The removal efficiency expressed in terms of % can be calculated using Eq. 2:

$$q_e = \frac{(C_0 - Ct) * V}{M} \tag{4.1}$$

$$\text{Removal Efficiency} = \frac{(C_0 - Ct) * 100}{C_0} \tag{4.2}$$

Recyclability of sorbents is an important aspect; therefore, desorption studies have to be carried out in conditions opposite to that of sorption; that is, if sorption occurs at slightly acidic or near-neutral conditions, then desorption will occur in highly acidic conditions. Desorption ratio $(\%)$ can be calculated as the ratio of the solute concentration sorbed to desorbed.

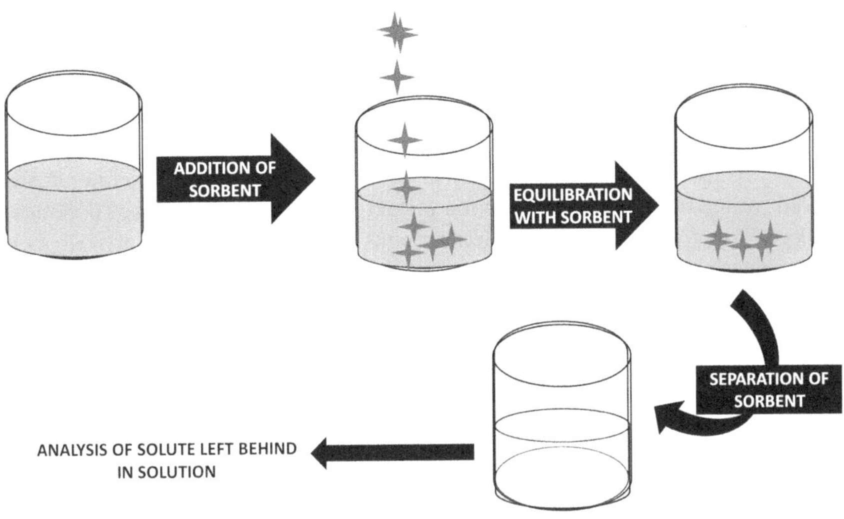

FIGURE 4.1 Schematic representation of sorption in batch mode

4.3 UNDERSTANDING SORPTION USING MODELS

When there is a contact between the sorbate and sorbent for a sufficient time period, an equilibrium is attained. The understanding of sorption equilibria is very important as it can give crucial information regarding the interactions of the different species with solid [6]. Sorption isotherms are curves obtained using mathematical expressions describing the equilibrium of sorbate distribution between two heterogeneous phases of solid and liquid at a constant temperature. It is used to calculate the maximum sorption capacity [amount of solute sorbed per unit mass of sorbent (mg/g)] and can give an idea of sorbent efficiency. However in general, it is seen that the efficiency is greatly affected by the nature of the sorbate, sorbent, pH, ionic strength, etc. [7–10]. Usually, it is quite easy to use linear regression analysis [11] to find the best suitable models as it computes the sorbate distribution and authenticates the postulations of the particular model. The inherent bias created by assuming a linear relation leads to erroneous results. The use of non-linear isotherm modelling is now gaining popularity to reduce the errors to a great extent. The adsorption isotherm of gas which was used traditionally is the plot of the amount adsorbed n_{ads} versus the relative pressure (P/P_0). It is reported that these isotherms could be categorized based on their shape [8, 12–16]. The independent domain theory [12] states that the relative vapour pressure should be larger during pore filling as compared to that of vacant pores to result in hysteresis. The different types of isotherms are as follows: (i) type I isotherm (convex upwards) is characterized by a horizontal plateau for high values of gas pressure and can be described by Langmuir equation, e.g. sorption of water vapour on zeolite or hydrogen on charcoal; (ii) type II isotherm (sigmoid curve) describes the monolayer adsorption on mesoporous materials at low pressure and multilayer sorption at high pressure near saturation with no hysteresis but a single inflection point (usually observed in microporous, non-porous, or disperse solids with >50 nm pore diameter). The adsorption of nitrogen on silica gel or a catalyst is an example of type II isotherm. (iii) Type III isotherm is a hyperbolic curve which concaves upwards. This arises when the interactions between two adsorbate molecules are more stronger than the adsorbate–adsorbent interactions. Adsorption of water on hydrophobic zeolites and activated carbon, bromine, and iodine on silica gel, and carbon tetrachloride adsorption on mesoporous gel follow type III isotherm. (iv) Type IV isotherms with two inflexion points describe the adsorption behaviour of mesoporous materials with pore condensation behaviour. This isotherm is obtained during the adsorption of benzene on iron oxide and on silica gel. (v) Type V isotherm with one inflexion point indicates the presence of mesopores formed due to pore condensation. The adsorption of water on activated carbon fibre follows this isotherm. (vi) Type VI isotherms with several inflexion points and steps indicate the formation of layers at low temperatures. Adsorption of noble gases on the surfaces of planar graphite, adsorption of butanol on aluminium silicate, and adsorption of CH_4 on MgO follow this type of isotherm.

A large number of isotherm models have been formulated to understand the isotherms, taking into account that the sorption is an equilibrium process (rates of sorption and desorption are equal) and also its thermodynamics. The conventional linear least-squares method is ideal for the determination of the most appropriate model and involves the goodness fit of a particular model with its regression correlation coefficient value close to 1 [17]. However, it is now seen that several restrictions related to the linearized isotherm expressions can result in a bias in the data [18, 19] as can be seen from the error analysis [20–22]. The performance of a sorbent can be estimated by the fitting of experimental data to different equilibrium and kinetic models. This helps in understanding the sorption mechanism and comparing the performance of different sorbents to help in the effective design of sorption procedures. It is essential to select the exact isotherm and calculate the parameters of this model to understand the behaviour of both the sorbate and sorbent and also the outcome of the overall process. The experimentally obtained data are subjected to different models to get an idea of the equilibrium and kinetic aspects of the sorption.

4.3.1 Equilibrium models

A large number of equilibrium models are available. It is seen that each of them have varying numbers of parameters. One-parameter model (Henry), two-parameter models (Langmuir, Freundlich, Dubinin–Radushkevich, Temkin, Hill–Deboer, Fowler–Guggenheim, Flory–Huggins, Halsey, Harkins–Jura, Jovanovic, Elovich, and Kiselev), three-parameter models (Hill, Redlich–Peterson, Sips, Langmuir–Freundlich, Fritz–Schlunder-III, Radke–Prausnitz-I, Radke–Prausnitz-II, Radke–Prausnitz-III, Toth, Khan, Koble–Corrigan, Jossens, Jovanovic–Freundlich, Brouers–Sotolongo, Vieth–Sladek, Unilan, Holl–Krich, and Langmuir–Jovanovic), four-parameter models (Fritz–Schlunder-IV, Baudu Weber–van Vliet, and Marczewski–Jaroniec), and five-parameter model (Fritz–Schlunder-V) can be used to analyse the experimental equilibrium sorption data. The meaningfulness of these models with respect to the sorption experimental data to envisage sorption is achieved using software to estimate not only the model parameters but also the values of the non-linear regression coefficient $\left(R^2\right)$, sum of squares due to error (SSE), and root mean squared error (RMSE). In the following section, the different models discussed are classified based on the number of parameters involved [23, 24].

1. *One-parameter model:* Henry's one-parameter isotherm model is the simplest relation between the amount of sorbate (substance sorbed on to the sorbent) and its partial pressure (for gas sorption) and is applied to low gas concentrations (sorbed molecules are far from their nearest neighbours). The equilibrium sorbate concentrations on the sorbent and in gas can be related by a simple equation (Eq. 3), where qe (mg/g) and C_e (mg/g) are concentrations on the sorbent and solution, respectively, at equilibrium, the amount of the sorbate at equilibrium, and k_{HE} is Henry's sorption constant. The linear isotherm can be used to describe the initial part of many practical isotherms. It is valid for low surface coverages, and the sorption energy is independent of the coverage (due to the absence of inhomogeneities on the sorbent surface):

$$q_e = K_{HE} C_e,\qquad(4.3)$$

2. *Two-parameter models:* In this section, the discussion is with respect to twelve models.
    (i) Langmuir isotherm model: Langmuir model presupposes that the sorbent surface is homogeneous (all sites have equal sorptive energy), sorption is localized (atoms/molecules are sorbed at definite localized sites), and each site can accommodate only one molecule/atom and sorption on one site does not affect sorption on the adjacent site (leading to monolayer coverage of the sorbent sites). Langmuir propositioned that the rate at which the sorbate molecules strike a sorbent surface is proportional to the product of the concentration of the sorbate and the fraction of the sorbent surface remaining uncovered by the sorbate. This model suggests that the rates of sorption and desorption are nearly equal at equilibrium and are proportional to the fractions of sorbent sites that are unoccupied and occupied, respectively, and are represented by Eq. 4 [wherein b is the Langmuir constant related to sorption capacity (mg/g), which correlates with the variation of the suitable area and porosity of the sorbent]. The capacity values computed using this model is useful for comparing the efficiency of various sorbents. The dimensionless constant, Langmuir separation factor R_L (Eq. 5), is used to predict whether the sorption is unfavourable $(R_L>1)$, linear $(R_L>1)$, favourable (0–1), or irreversible $(R_L = 0)$.

$$\frac{C_e}{q_e} = \frac{1}{Q^o * b} + \frac{C_e}{Q^o}\qquad(4.4)$$

$$R_L = \frac{1}{1 + C_e * b}\qquad(4.5)$$

(ii) Freundlich isotherm model: The Freundlich model gives an idea of heterogeneity of sorbent surface and is given by Eq. 6, wherein a_F is Freundlich sorption capacity (L mg/g) and n_F is sorption intensity. The value of $1/n_F$ is in the range 0–1 for favourable sorption, with the value close to zero indicates more heterogeneous nature of sorption sites.

$$\ln q_e = \ln K_F + \frac{1}{n}\ln C_e \qquad (4.6)$$

(iii) Dubinin–Radushkevich (D-R) isotherm model: This empirical model, applied originally for vapour sorption onto microporous solids, assumes a multilayer sorption involving van Der Waals forces leading to a pore-filling mechanism. It is used to estimate the porosity and free energy of sorption. This isotherm is applicable for intermediary values of sorbate concentrations and can be used to differentiate between physisorption and chemisorption processes based on the energy calculated $(\text{E} < 8\,(\text{kJ/mol})$ for physisorption; $\text{E} = 8 - 16\,(\text{kJ/mol})$ for chemisorption). D-R isotherm is dependent on the temperature of sorption. The model is represented by Eq. 7, wherein the degree of heterogeneity $\beta\,(0 < \beta < 1)\,\beta$ and the Polanyi potential ε given by RT/ln $(1+1/C_e)$ can be calculated. The apparent or free mean energy (kJ mol) of the sorption/sorbate molecule Es for sorption from infinity in the solution to the solid surface can be computed by Eq. 8.

$$\ln q_e = -\beta\varepsilon^2 + \ln Q \qquad (4.7)$$

$$E_s = \frac{1}{\sqrt{2}\times\beta} \qquad (4.8)$$

(iv) Temkin isotherm model: This model was originally used for gas sorption in an intermediate concentration range and gives an idea of the effect of sorbate–sorbate interactions on the overall sorption process. According to this model, the heat of sorption decreases linearly with the increase in coverage, and sorption is characterized by the uniform distribution of binding energies. The model is represented in Eq. 9, wherein RT/bT correlates the heat of sorption, while the equilibrium binding constant A_T (L/mg) represents the maximum binding energy.

$$q_e = \left(\frac{RT}{b}\right)*\ln A + \left(\frac{RT}{b}\right)*\ln C_e \qquad (4.9)$$

(v) Hill–Deboer isotherm model: This model explains the mobile sorption and lateral interactions between sorbed molecules and is given in Eq. 10. K_1 (L/mg) and K_2 (kJ/mol) are Hill–Deboer and energetic constants, respectively. The positive value of K_2 indicates the attraction between sorbed species, while the negative value indicates repulsion and a zero value of K_2 indicates no interaction between sorbed molecules, and it reduces to the Volmer equation.

$$\ln\left(\frac{C_e(1-\theta)}{\theta}\right) - \frac{\theta}{1-\theta} = -\ln K_1 - \frac{K_2\theta}{RT} \tag{4.10}$$

(vi) Fowler–Guggenheim isotherm model: This model gives information on the lateral interaction of sorbed molecules. It indicates that the heat of sorption varies in a linear relationship with loading. The expression for the model is given in Eq. 11. The positive and negative values of W are indicative of the interaction and repulsion, respectively, between the sorbed molecules. When there is no interaction, the value becomes zero, and the model is reduced to the Langmuir model. This model also relates the interaction between the sorbed molecules and the heat of sorption. The attraction between sorbed molecules leads to an increase in the heat of sorption, while repulsion decreases the heat of adsorption.

$$\ln\left(\frac{C_e(1-\theta)}{\theta}\right) - \frac{\theta}{1-\theta} = -\ln K_{FC} + \frac{2W*\theta}{RT} \tag{4.11}$$

(vii) Flory–Huggins isotherm model: This model, given in Eq. 12, is useful in estimating the degree of surface coverage. This model also helps in understanding whether the sorption process is spontaneous. In Eq. 12, θ is the degree of surface coverage $(1Ce/Co)$, K_{FH} is the Flory–Huggins model constant, and n is the number of adsorbate occupying a given site. The free energy of the sorption is calculated using K_{FH}.

$$\log\frac{\theta}{C_e} = -\ln K_{FH} + n\log(1-\theta) \tag{4.12}$$

(viii) Halsey isotherm model: This model gives an idea of multilayer sorption in a space quite far away from the sorbent surface. The fitting of the experimental data to this equation proves the heterogeneous distribution of active sites and also multilayer sorption. The mathematical expression of this model is given in Eq. 13:

$$\ln q_e = \frac{1}{n}*\ln K - \frac{1}{n}*\ln C_e \tag{4.13}$$

(ix) Harkins–Jura isotherm model: This model explains the multilayer sorption on the surface of sorbents with heterogeneous pore distribution, and the mathematical expression is given in Eq. 14. The plot of 1/qe2 vs log Ce can give the values of the constants A and B.

$$\frac{1}{q_e^2} = \frac{B}{A} - \left(\frac{1}{A}\right)*\log C_e \tag{4.14}$$

(x) Jovanovic isotherm model: This model is similar to that of the Langmuir model with respect to monolayer localized sorption, but there is also a possible interaction or contact between the sorbed and desorbed molecules. The mathematical expression is given in Eq. 15. At high concentrations of the sorbate, it becomes the Langmuir isotherm.

$$\ln q_e = \ln q_{max} - K_f C_e \qquad (4.15)$$

(xi) Elovich isotherm model: The model assumes multilayer adsorption and is used for describing chemisorption on highly heterogeneous sorbents. The Elovich isotherm model is given in Eq. 16, wherein q_m is the Elovich adsorption capacity and K_e is the Elovich constant obtained from the plots of $\ln\left(q_e/C_e\right)$ vs q_e.

$$\ln\frac{q_e}{C_e} = \ln\left(Ke*q_m\right) - \frac{q_e}{q_m} \qquad (4.16)$$

(xii) Kiselev isotherm model: This model explains the localized sorbate monolayer formation on sorbents. This model is applicable when surface coverage is larger than 0.68. The Kiselev adsorption isotherm model is given in Eq. 17, wherein K1 and Kn are the Kiselev equilibrium constant and complex (between adsorbate molecules) formation constant, respectively.

$$\frac{1}{C_e\left(1-\theta\right)} = \frac{K_1}{\theta} + K_1 K_n \qquad (4.17)$$

3. *Three-parameter models:* The models containing three parameters to explain the mechanism of adsorption are discussed using sixteen models, viz., Hill, Redlich–Peterson, Sips, Langmuir–Freundlich, Fritz–Schlunder-III, Radke–Prausnitz, Toth, Khan, Koble–Corrigan, Jossens, Jovanovic–Freundlich, Brouers–Sotolongo, Vieth–Sladek, Unilan, Holl–Krich, and Langmuir–Jovanovic.

(i) Hill isotherm model: The model is developed to illustrate the sorption on homogeneous substrates. This model assumes that the sorbed molecule on one site would influence the sorption on other sites of the sorbent via a cooperative mechanism. The model is given in Eq. 18. The values of n_H give an idea of the nature of cooperation; values are greater, equal to, or less than 1 for positive cooperation, non-cooperation, and negative cooperation, respectively, in binding.

$$q_{eq} = \frac{q_{max} C_{eq}^{nH}}{K_H + C_{eq}^{nH}} \qquad (4.18)$$

(ii) Redlich–Peterson (R-P) isotherm model: The R-P isotherm model has characteristics of both Langmuir and Freundlich isotherms. Therefore, the mechanism is an amalgamation of both and is not ideally monolayer sorption. The R-P isotherm model is given in Eq. 19. This model is applicable over a wide concentration range and applicable to both homogeneous and heterogeneous systems. At high liquid-phase concentrations of the sorbate, this model reduces to Freundlich model and towards Henrys law at low concentration. The value of the exponent is in the range of 0–1; at 1, the model approaches the Langmuir model, while at 0, it approaches the Freundlich model.

$$q_{eq} = \frac{A_{RP} C_{eq}}{1 + B_{RP} C_{eq}^{\beta}} \tag{4.19}$$

(iii) Sips isotherm model: This model deals with the localized sorption of the sorbate without any interactions with the sorbent. This model is a combination of Langmuir and Freundlich models and is used for heterogeneous adsorption systems. At low sorbate concentrations, the model changes to the Freundlich model, and at high concentrations, it predicts a monolayer sorption that is characteristic of the Langmuir model. Sips adsorption isotherm model is given in Eq. 20.

$$q_{eq} = \frac{q_m K_s C_{eq}^{\beta s}}{1 + K_s C_{eq}^{\beta s}} \tag{4.20}$$

(iv) Langmuir–Freundlich isotherm model: This model describes the sorption in heterogeneous surfaces. At low sorbate concentration, the model approaches the Freundlich model, while at high concentrations, it approaches the Langmuir isotherm model. The model is represented in Eq. 21, wherein m_{LF} is the heterogeneous parameter with a value between 0 and 1. The decrease in surface heterogeneity results in an increase in m_{LF}, and when it equals 1, it becomes the Langmuir model.

$$q_{eq} = \frac{q_{max} \left(K_{LF} C_{eq} \right)^{m_{LF}}}{1 + \left(K_{LF} C_{eq} \right)^{m_{LF}}} \tag{4.21}$$

(v) Fritz–Schlunder-III isotherm model: This three-parameter isotherm model is developed to fit over an extensive range of experimental results. The sorption is expressed in Eq. 22, wherein the values of m_{FS3} of 1 or 0 leads to the change of the Fritz–Schlunder-III model to Langmuir and Freundlich models, respectively.

$$q_{eq} = \frac{q_{max} K_{FS3} C_{eq}}{1 + q_{max} C_{eq}^{m_{FS3}}} \tag{4.22}$$

(vi) Radke–Prausnitz isotherm model: This is a preferred model for low sorbate concentration and is applicable over a wide concentration range. At low sorbate concentration, the model reduces to the linear Henry law, and at higher sorbate concentration, it tends to become the Freundlich model. When the value of the Radke–Prausnitz exponent (m_{RaP3}) is zero, this model becomes the Langmuir model. The mathematical expressions of the model are given in Eqs. 23–25. If the values of both m_{RaP1} and m_{RaP2} are 1, the Radke–Prausnitz-I, II models reduce to the Langmuir model. At low concentrations, the models become the Henry model, but for high sorbate concentration, the Radke–Prausnitz-I and II models become the Freundlich model. The Radke–prausnitz-III equation reduces to Henry and Langmuir models when the values of m_{RaP3} are 1 and 0, respectively.

$$\text{Model 1} \quad q_{eq} = \frac{q_{max} K_{RaP1} C_{eq}}{\left[1 + K_{RP1} C_{eq}\right]^{m_{RaP1}}} \tag{4.23}$$

$$\text{Model 2} \quad q_{eq} = \frac{q_{max} K_{RaP2} C_{eq}}{1 + K_{RP2} C_{eq}{}^{m_{RaP2}}} \tag{4.24}$$

$$\text{Model 3} \quad q_{eq} = \frac{q_{max} K_{RaP3} C_{eq}^{m_{RaP3}}}{1 + K_{RP3} C_{eq}{}^{m_{RaP3-1}}} \tag{4.25}$$

(vii) Toth isotherm model: This model is developed to describe the heterogeneous sorption systems at both low and high sorbate concentrations. This model can be considered as a modified form of the Langmuir model, with an aim to reduce the error between experimental and computed data and is given in Eq. 26; at n = 1, this equation becomes the Langmuir isotherm equation, indicating that the process approaches a homogeneous surface, and values deviating from 1 indicates a heterogeneous surface. Thus, the parameter n is an attribute of heterogeneity of sorption and is suited for modelling multilayer and heterogeneous sorption.

$$q_{eq} = \frac{q_{max} C_{eq}}{\left(\dfrac{1}{K_T} + C_{eq}^{n_r}\right)^{\frac{1}{m_r}}} \tag{4.26}$$

(viii) Khan isotherm model: This model is developed for the sorption of bi-adsorbates, and it reaches Freundlich and Langmuir isotherms at the two concentration levels. The mathematical expression is given in Eq. 27; at $a_K = 1$, the model approaches the Langmuir isotherm, and at higher concentration values, it becomes the Freundlich isotherm model.

$$q_{eq} = \frac{q_m b_K C_{eq}}{\left(1 + b_K C_{eq}\right)^{aK}} \tag{4.27}$$

(ix) Koble–Corrigan isotherm model: This model incorporates both Langmuir and Freundlich isotherms and is similar to the Sips isotherm model. This is expressed as given in Eq. 28. This model reduces to the Freundlich model at high sorbate concentrations and is valid for n ≥ 1 and becomes incapable of defining the experimental data despite high concentration at 1 ≤ n.

$$q_{eq} = \frac{A_{KC} C_{eq}^{m_{KC}}}{1 + B_{KC} C_{eq}^{m_{KC}}} \tag{4.28}$$

(x) Jossens isotherm model: The model is based on the energy distribution of sorbate–sorbent interactions at sorption sites. The assumption made in this model is that the sorbent has a heterogeneous surface with respect to the sorbate interactions. At low concentrations, this model is reduced to Henry's law. The mathematical expression is given in Eq. 29, wherein J is Henry's constant at low capacities, b_J the Jossens isotherm constant characteristic of the sorbent, irrespective of temperature.

$$q_{eq} = \frac{K_J C_{eq}}{1 + JC_{eq}^{bJ}} \tag{4.29}$$

(xi) Jovanovic–Freundlich isotherm model: This model is developed to describe single-component sorption equilibrium on heterogeneous surfaces. The main assumption is that the rate of decrease in the fraction of the sorbent surface unoccupied by the sorbate molecules is proportional to a certain power of the partial pressure of the sorbate. For homogenous sorbent surface, this model reduces to Jovanovic. At low and high pressures, the model becomes Freundlich and Langmuir, respectively. Similar to the Jovanovic model, this model also takes into account the contact between sorbed and desorbed molecules. The mathematical expression is given in Eq. 30.

$$q_{eq} = q_{max}\left[1 - e^{-\left(K_{JF} C_{eq}^{nJF}\right)}\right] \tag{4.30}$$

(xii) Brouers–Sotolongo isotherm model: This isotherm is in the form of a deformed exponential function for sorption onto a heterogeneous surface, and it appears to be the extension of the Langmuir isotherm to irregular sorbent surfaces. The main assumption is that the surface consists of clusters of active sites of equal energies. The mathematical expression of this model is given in Eq. 31. The exponential term is related to the sorption energy distribution and gives the value energy of heterogeneity at a particular temperature.

$$q_{eq} = q \left[1 - e^{\left(-K_{HS} C_{eq}^{\,nHS} \right)} \right]_{max} \tag{4.31}$$

(xiii) Vieth–Sladek isotherm model: This model integrates two discrete segments for the calculation of diffusion rates in sorbents. The first one is defined by a linear section (Henry's law), and the second one is a non-linear section (Langmuir isotherm). The linear section correlates to the physisorption of gas molecules onto amorphous sorbent surfaces, while the non-linear section explains the adherence of gas molecules on the porous sorbent sites. The mathematical expression is given by Eq. 32.

$$q_{eq} = K_{VS} C_{eq} + \frac{q_{max} \beta_{VS} C_{eq}}{1 + \beta_{VS} C_{eq}} \tag{4.32}$$

(xiv) Unilan isotherm model: This model is applicable for Langmuir isotherm and uniform energy distribution. This equation becomes Henry's law at very low sorbate concentration and is applied to gas sorption on a heterogeneous sorbent surface. This model is given in Eq. 33. The value of the model exponent U is a measure of heterogeneity; the higher the value, the more the heterogeneity. If $U = 0$, the isotherm model becomes the classical Langmuir model as energy distribution is zero.

$$q_{eq} = \frac{q_{max}}{2\beta_U} \, Ln \left[\frac{1 + K_U C_{eq} e^{\beta U}}{1 + K_U C_{eq} e^{-\beta U}} \right] \tag{4.33}$$

(xv) Holl–Krich isotherm model: This model is a modified version of Langmuir isotherm, and at low concentrations, it becomes Freundlich isotherm. The capacity saturates more slowly as compared to Langmuir isotherm at high concentrations. The expression is given in Eq. 34:

$$q_{eq} = \frac{q_{max} K_{HK} C_{eq}^{n_{HK}}}{1 + K_{HK} C_{eq}^{n_{HK}}} \tag{4.34}$$

(xvi) Langmuir–Jovanovic isotherm model: This pragmatic model is an amalgamation of Langmuir and Jovanovic isotherms and is given by Eq. 35.

$$q_{eq} = \frac{q_{max} C_{eq} \left[1 - e^{K_{LJ} C_{eq}^{n_{LJ}}} \right]}{1 + C_{eq}} \tag{4.35}$$

4. *Four-parameter models:*

  (i)   Fritz–Schlunder-IV isotherm model: This is a four-parameter model
         with collective facets of Langmuir–Freundlich isotherms and is given
         by Eq. 36. This isotherm is valid when the values of α_{FS5} and β_{FS5} are
         ≤ 1. At high sorbate concentrations, the Fritz–Schlunder-IV isotherm
         becomes Freundlich equation, and when α_{FS5} and β_{FS5} are $= 1$, this
         equation reduces to a Langmuir isotherm. At high sorbate concentra-
         tions, this isotherm becomes a Freundlich isotherm.

$$q_{eq} = \frac{A_{FS5} C_{eq}^{\alpha_{FS5}}}{1 + B_{FS5} C_{eq}^{\beta_{FS5}}} \qquad (4.36)$$

  (ii)   Baudu isotherm model: This has been developed mainly due to the
         discrepancy in the calculation of Langmuir constant and coefficient
         for a wide range of concentrations. It can be considered to be the
         altered form of the Langmuir isotherm and is given in Eq. 37. This
         model is only applicable in the range of (1+x+y)<1 and (1+x)<1. For
         lower surface coverage, the Baudu model reduces to the Freundlich
         equation and is given in Eq. 38.

$$q_{eq} = \frac{q_{maxbo} C_{eq}^{(1+x+y)}}{1 + b_o C_{eq}^{(1+x)}} \qquad (4.37)$$

$$q_{eq} = \frac{q_{mo\,bo} C_{eq}^{(1+x+y)}}{1 + b_o} \qquad (4.38)$$

  (iii) Weber–van Vliet isotherm model: This model is another four-parameter
         model and is given in Eq. 39. The isotherm parameters P1, P2, P3, and
         P4 can be defined by multiple non-linear curve fitting techniques using
         minimization of sum of square of residual.

$$C_{eq} = P_1 q_{eq}^{\left(P_2 q_{eq}^{P3} + P4\right)} \qquad (4.39)$$

  (iv)   Marczewski–Jaroniec isotherm model: This model resembles the
         Langmuir model and was developed on the belief that localized sorp-
         tion showed Langmuir coverage and there is a distribution of sorption
         energies. The model is given in Eq. 40, wherein K_{MJ} and n_{MJ} describe
         the spreading of distribution in the path of higher and lesser sorption
         energies, respectively. When $n_{MJ} = K_{MJ}$, the isotherm reduces to the
         Langmuir–Freundlich model, and when both are equal to 1, it reduces
         to the Langmuir isotherm.

$$q_{eq} = q_{max} \left[\frac{(K_{MJ} C_{eq})^{n_{MJ}}}{1 + (K_{MJ} C_{eq})^{n_{MJ}}} \right] \qquad (4.40)$$

5. *Five-parameter models:* The high parameter model will provide very lucid information on sorption mechanism under equilibrium conditions. The Fritz–Schlunder-V isotherm model is developed with the aim of replicating the variations precisely over a wide concentration range. The isotherm model is given in Eq. 41.

$$q_{eq} = \frac{q_{max} K_{1FS5} C_{eq}^{\alpha FS5}}{1 + K_{2FS5} C_{eq}^{\beta FS5}} \tag{4.41}$$

There are many models to which the experimentally obtained sorption data can be fitted to. However, it is the curve with the best fit that predicts the optimum model. The next important aspect is that the parameters obtained from the model should be realistic.

4.3.2 KINETIC MODELS

Sorption kinetics portrays the progression of the sorption process with time, till its equilibrium is reached. This presents information regarding the probable sorption mechanism. There are two kinetic models, namely "pseudo-first-order" and "pseudo-second-order" rate expressions, and two sorption mechanism models, namely, Webber–Morris intraparticle and Boyd's film diffusion models.

(i) Pseudo-first-order kinetic model: Lagergren or pseudo-first-order equation is the earliest equation describing the sorption rate and is given in Eq. 42, wherein q_t and q_e are the amounts of sorbate sorbed at time t and at equilibrium, respectively, and k_{ads} is the rate constant or time-scaling parameter. The straight line plot of log ($q_e - q_t$) vs t is used for the calculation of K_{ads} and q_e from the slope and intercept, respectively. Accurate determination of q_e is a difficult task because, in many sorbate–sorbent interactions, chemisorption becomes very slow after the initial fast response, and it is difficult to ascertain whether equilibrium can be reached. It is usually observed that sorption processes normally follow the Lagergren or pseudo-first-order model only for the initial 20–30 minutes of interaction. The value of k_{ads} decides how fast the equilibrium can be reached in the system. The experimental studies show that the value of k_{ads} can be either dependent on the initial concentration of the sorbate or independent of the experimental conditions.

$$\log(q_e - q_t) = \log q_e - \frac{k_{ads} \times t}{2.303} \tag{4.42}$$

(ii) Pseudo-second-order kinetic model: This deals with the situation when the rate of direct sorption/desorption process (considered as a chemical reaction) controls the overall sorption kinetics. The model developed by Ho and McKay is expressed in Eq. 43, wherein k_{2ads} (rate constant) and q_e (amount sorbed at equilibrium) can be easily computed from the linear plot of t/q_t vs t. It is understood that the value of k_{2ads} has a strong dependence on the applied operating conditions, such as the initial solute concentration, pH of the solution, and temperature.

$$\frac{t}{q_t} = \frac{1}{k_{2ads} \times q_e^2} + \frac{t}{q_e}$$
(4.43)

(iii) Webber–Morris intraparticle diffusion model: The intraparticle diffusion process rivets around ion migration into the internal surface of the sorbent due to the presence of pores. The boundary layer diffusion and the surface sorption rates (characteristic of sorbents) affect intraparticle diffusion. The model is articulated by the mathematical expression given in Eq. 44. In this equation, k_d is the intraparticle diffusion rate, and I (mg/g) is a constant indicating the size of the boundary layer, which are the slope and intercept, respectively, of the linear plot of q_t vs $t^{1/2}$.

$$q_t = k_d t^{0.5} + I$$
(4.44)

(iv) Boyd diffusion model: This model is used to ascertain if sorption progresses via film diffusion or intraparticle diffusion mechanism. The mathematical expression of Boyd's equation is given in Eq. 45, in which $F = q_t/q_e$, where q_t and q_e are the amounts of metal ion sorbed on the surface (µg/g) at time t and at equilibrium, respectively. The value of B was used to calculate the effective diffusion coefficients D_i (cm^2/s) using Eq. 46, wherein r is the radius of the sorbent. The linearity of the Bt *vs.* t plot provides useful information to distinguish between the film and intraparticle diffusion. A straight line passing through the origin is indicative of intraparticle diffusion-controlled sorption.

$$B * t = -\ln(1 - F) - 0.4977$$
(4.45)

$$B = D_i * \left(\frac{\pi}{r}\right)^2$$
(4.46)

4.4 APPLICATIONS

Sorption is proving to be a lucrative protocol for the eradication of toxic species. The main aspect of using any sorbent is that it itself should not impart any further aberration to the ground water system. Thus, it becomes important that the sorbent material be of great stability.

The use of charcoal was first reported for the elimination of taste and odour from contaminated water [25, 26]. Activated carbon can be prepared from a variety of materials (bones, coconut coir, human hair, paper, tree bark, sunflower seeds, corn cobs, tea leaves, coffee beans, fish, municipal waste, etc), and the functional groups present are governed by the nature of the precursor and activation method [27–31]. The activation process creates reproducible porous structures with mesopores, micropores, and ultra-micropores having a large surface area. Activated carbon is a very effective sorbent due to its high surface area, but it is expensive and is also not suited for hydrophilic species. Furthermore, the regeneration is even more costly, and loss of sorbent is encountered. Therefore, various other materials have been evaluated for their potential use as sorbents for remediation. Ion exchange resins appear to be attractive due to their selectivity and ease of regeneration but find limited use in water remediation applications due to their high cost and change in structure due to pH. As it is understood that sorption is primarily a surface phenomenon, surface area is expected to play a role. Hence, it is of great interest to examine the sorption efficiency of particles with reduced size. Nanoparticles (NPs; size 11–100 nm) make up a new realm of matter wherein physical and chemical properties are drastically different and sensitive to size changes, leading to a wide range of applications [32]. Residual surface hydroxides can also add to the rich surface chemistry of metal oxides, including sorption characteristics [33]. The use of nanomaterials for remediation of waste water is gaining a lot of attention [34, 35]. In order to enhance the uptake capacity and obtain selective sorption performance, modification of sorbents is an attractive option. This can be carried out by either surface functionalization using specific ligands or impregnation of the ligands within the sorbent material. There are large numbers of studies on the use of different materials for remediation of various kinds of pollutants. As expected, the work reported in the literature in the field of sorption is immeasurable, and it is not possible to include all work in the present chapter. Therefore, some specific examples relating to certain types of sorbents or based on some common pollutants have been given to provide an understanding of the application of the sorption process.

Transition metal nanooxides such as ferrites, mixed ferrites, and manganese oxide were found to take up lead ions without any surface modification. It has been observed that the nature of the metal oxide played a key role in the characteristics of the nanoparticles and subsequently on the sorption process [36]. Nanocrystalline manganese oxide with a small particle size and high surface was found to sorb various metal ions at different pH values with different kinetics [37]. Zinc oxide, a multifunctional nanomaterial, was used for sorption of different species. The use of zinc oxide as sorbent is also quite interesting. The charge on its surface relied on the synthesis protocol [38–40]. It was seen that pyrolysis and gel combustion methods imparted a positive surface charge, while the co-precipitation method made the surface negative. It is more interesting to note that the ZnO NPs obtained by the three syntheses protocols had identical structures. Thus, the zinc oxide applications vary according to the surface charge, and ZnO NPs (obtained from pyrolysis and gel combustion) were found to be excellent sorbents for anionic chromate species. The mechanism of sorption proposed was a combination of electrostatic interaction, pore filling, and complex formation. Negatively charged ZnO NPs (obtained from the co-precipitation method) showed predilection to $Cu(II)$ ions. In order to introduce selectivity in the sorption ability of ZnO NPs with respect to lead and mercury, surface modification or functionalization by the introduction of appropriate ligand during synthesis was applied [41]. X-ray diffraction (XRD) shows that ligand introduction does not alter the structure or average size of ZNO NPs, but crystallinity was affected.

Fluoride is considered to be one of the most common pollutants, and WHO has specified a limit of 1.5 mg/L in drinking water. Iron-based composites with polypyrrole (PPy)/FE_3O_4 showed excellent fluoride sorption efficiency [42, 43]. Amorphous FE/Al mixed hydroxides with a high surface area showed that fluoride formed new complexes on the sorbent surface [44, 45]. The sorption was adversely affected by phosphate, sulphate, and arsenate [45]. The granulated mixture of Fe/Al–Ce nano-sorbent sprayed onto glass beads as column materials for fluidized bed was used for fluoride removal [46]. The complex formation with Fe(III)-modified natural stilbite zeolite was liable for fluoride removal [47]. Calcium-based sorbents showed that the mechanism of removal was a combination of both surface sorption and precipitation, and the sorption was affected by surface area [48–53]. Al–Ce hybrid sorbents prepared by co-precipitation were found to contain nanoparticle aggregates amidst an amorphous structure as revealed by SEM and XRD studies and were found to have an uptake capacity of 27.5 mg/g for fluoride. Magnesium-based sorbents were found to show excellent sorption efficiency for the removal of fluoride from aqueous solutions [54–57]. Nanocrystalline magnesia–alumina mixed oxide synthesized by the combustion method could remove fluoride at a near-neutral pH (~6) within 1 hour [58]. Carbon-based composites [59–62] obtained from biomaterials like *Murrayakoeingii* (curry leaf seeds) [63] were found to have excellent sorption capacity of fluoride. Carbon-based materials have excellent defluoridation capacity due to their properties, but the need to make them more cost-effective and environmental friendly is a challenge to and interest of many researchers [64]. Non-conventional sorbents based on cement [65–67] prove useful for the removal of fluoride. The use of hydroxyapatite and its modified forms as sorbents for water defluoridation has been reported [68]. Nanocomposites like cellulose/hydroxyapatite [69] and aluminium-modified hydroxyapatite (Al-HAP) [70] show excellent fluoride removal capacity. Zirconium-based sorbents are very attractive due to their high efficiency [71–76]. Activated alumina (AA) without and with modification have been found to show excellent deflouridation property [77–82]. Unmodified alumina can be used only for pH below 6 due to its leaching. Therefore, modification can reduce the leaching and show application in a wider range of pH. Lanthanum(III) and ytterbium(III) impregnated on alumina have shown very promising results for defluoridation of water. Chromium is another pollutant which is of great concern. However, in case of chromium, toxicity depends on the specific form of chromium. Trivalent chromium, or chromium(III), is an essential micronutrient, whereas hexavalent chromium, or chromium(VI), is unequivocally toxic, leading to haemolysis and, ultimately, kidney and liver failure. The maximum allowable concentration of Cr(VI) according to WHO is 0.05 mg/L in drinking water. Commercially available activated carbons (bare or modified) have been used for the sorption of chromium [83–87]. It is seen that sorption results in the reduction of the hexavalent form to the trivalent form. The sorption capacity of the oxidized carbon was greater than that of the original carbon due to increased surface charge of the ionized oxygen atoms (resulting in an increase in the electrostatic attraction between the carbon surface and metal cations). Low-cost sorbents such as coir pith [88], sawdust [89, 90], straw [91], and leaf mould [92] have been used for chromium removal from water. Carboxylic, phenolic, and protonated amide groups help in the removal of hexavalent chromium; therefore, pretreatment with these groups resulted in a positive effect. The nature of shells of different nuts like coconut (CS), almond (AS), ground nut (GS), and walnut (WS) was found to affect sorption, with walnut being most efficient and coconut being the least. Although the mechanism is chemisorption, it is not clear as to the nature of the species being sorbed (anionic chromate or cationic Cr(III) formed by reduction) [92, 93]. Arsenic like chromium is a pernicious species whose toxicity is dependent on its valency. The period of exposure leads to various health issues from vomiting, abdominal pain, to more dangerous problems of skin darkening and cancer. According to WHO, the recommended level in water is less than 50 µg/L. Thus the removal of arsenic from water media becomes very crucial. Activated carbons have been extensively used for arsenic sorption [94] from aqueous solutions. It is seen that modification of activated carbon showed a marked enhancement of sorption efficiency [95, 96]. Agricultural waste [97] or its biochar material [98], bone char [99], and red mud [100, 101] have been used for arsenic sorption. Radionuclides are a very important class of metal ions which need to be removed from water bodies. Borosilicate glasses are extensively used in the immobilization of radioactive waste by the process of vitrification [102]. Thus, it was of interest to explore the room temperature and sorption capacity of these materials. Boroaluminosilicate glass could be used for the selective uptake of thorium from a mixture containing uranium [103]. The pH of the solution and composition of the glass have a profound consequence on selectivity. The use of glass as room temperature sorbents for transition metal ions has been studied [104]. It was seen that glasses in general can behave as ion exchangers due to their structural features, and to enhance selectivity and sorption kinetics, modification proved to be effectual. The use of tri-n-octylphosphine oxide (TOPO) and 8-hydroxyquinoline (oxine) was found to enhance the selectivity and kinetics of the removal of uranyl ions from its mixture [105]. The success of utilization of ligand-incorporated glass sorbents paved the way for the work on the use of AmberLite XAD 4 beads infused with n-benzoyl-n-phenylhydroxylamine (n-BPHA) for the selective and highly efficient removal of Th(IV) from its mixture [106]. The modified sorbent was characterized to confirm the presence of n-BPHA within the macroreticular resin structure. The uptake was nearly complete and highly selective in the pH range of 3–7.5, and selective capacity was calculated to be 500 mg/g. Polyaniline (PANI) synthesized by the chemical oxidation of aniline was studied for the uptake of uranyl ions [107]. This system was applied for the preconcentration of uranyl ions from seawater. Highly crystalline monoclinic antimony phosphate synthesized as both particles (size ~ 14 nm) and nanoribbons (length 500–700 nm with particles of size 1–5 nm) was found to show increased sorption efficiency for uranyl ions as compared to other metal ions [108]. Nanocrystalline manganese oxide (prepared by the hydrolysis of $KMnO_4$) with size of 8 nm and surface area of 145 m^2/g was found to show high uptake efficiency for most of the metal ions. However, it was observed that equilibration pH and time were parameters which could be used to achieve separation [109]. Functionalization of multi-walled carbon nanotubes (MWCNTs) in different ways was found to enhance the removal of Th^{4+} ions in aqueous solution [110]. The pH of a solution and nature of functionalization had a strong effect on the sorption efficiency.

Dyes are another class of contaminants that have attracted many researchers. Untreated wood and coal were found to be good sorbents for different dyes [111–113]. Peat, a natural sorbent, has been proven to be useful for the removal of dyes [114,

115]. Chitin and chitosan can be used as flakes, gels, beads, or fibres for the successful removal of dyes [116]. It is reported that a gamut of clay materials (bentonite, common clay, sepiolite, fire clay, fuller's earth, and kaolin) [117, 118] and industrial waste or by products [119, 120] can be used as sorbents for dyes. The mechanism of removal was attributed to the electrostatic attraction between the negative clay surface and dye cations. Boroaluminosilicate and barium borosilicate glasses have also been used for the removal of dyes like rhodamine 6G (R6G) and methylene blue (MB) [121, 122]. It has been observed that upon sorption of dyes, the glass matrix is not affected and the dye characteristics are also not altered. The glass structure contains a random arrangement of Si-O groups and thus the Si-O group is known as network former. Network modifiers tend to change the arrangement in various ways. Hence the glass composition can have different ratios of both the network former and modifier groups. In barium borosilicate glass, the oxides of alkali and alkaline metal ions are the modifiers while silica is the network former. Thus for the same ratio of network:modifier, it was seen that addition of barium oxide as compared to sodium oxide as modifier enhanced the dye uptake efficiency. This is attributed to the variation in the structural aspects between the two glasses, which have been compared using NMR studies. Novel nano-sorbents with exotic structures have been found to have excellent capacity for dyes with very fast kinetics. $MnWO_4^{\circ}$ and $MnMoO_4^{\circ}$ nanoparticles synthesized using a sonochemical technique had a high surface area and could remove cationic dyes within a very short time period [123]. It was also seen that $MnWO_4$ could sorb $Cu(II)$ ions. The thermal regeneration allowed the reuse of these novel nanomaterials for ten consecutive cycles. It was seen that the sonochemically synthesized barium molybdate $(BaMoO_4)$ nanoparticles demonstrated excellent sorption efficiency with respect to dyes [124]. The excellent property of $BaWO_4$ [125] as a sorbent was discovered serendipitously when the photocatalytic property of this material was evaluated. It was observed that the sorption of dyes was very fast and nearly complete. The tungstate and molybdate nanoparticles of Sr and Y synthesized by the sonochemical method could be used for the removal of cationic dyes and metal ions (non-radioactive and radioactive) [126]. $SrWO_4$ could selectively remove rhodamine dye from its mixture containing MB. It has been observed that the sorption efficiency correlated well with the surface charge and crystal structure.

A composite is made up of two or more materials that can be engineered or are naturally occurring. The two components have significantly different physical or chemical properties, but a composite has a combination of these properties. A simple system of iron oxide–silica composites shows that both the individual components have excellent applications. The two components have opposite surface charges, and therefore, the study of the sorption efficiency of the nanocomposite could prove useful to understand the sorption mechanism of metal ions and cationic dyes. Iron oxide–silica nanocomposites showed the uptake of toxic species like inorganic metal ions $\left(UO_2^{2+}, Zn^{2+}, Cu^{2+}, Ni^{2+}, Co^{2+}\right)$ and cationic dyes (MB and R6G) from aqueous streams at room temperature [127]. The modelling of sorption data was carried out to understand the sorption mechanism and understand the selectivity amongst the species. The surface modification of the sorbent enhanced its selectivity with respect to uranyl ion sorption. When the composites consist of natural products, they are termed as green composites [128], which can be defined as "materials composed wholly or in part of constituents which come ultimately from renewable sources". Green composites combine plant fibres with natural resins to create natural composite materials. Biodegradable polymers, which are the second component used for the production of green composites, can be classified into natural and synthetic biopolymers. Natural biopolymers can be further classified into two, namely, those obtained from the extraction of biomass and the others that are produced by organisms. Biopolymers obtained from extraction procedures are polysaccharides (like thermoplastic starch, chitin and chitosan, cellulose, and lignocellulose), polypeptides (like collagen, gelatin, corn, wheat, soy protein, caesin, and whey protein), and lipids (usually cross-linked). Biopolymers obtained from microorganisms are further classified based on the route of production, namely, microbial (polyhydroxy alkanoate, polyhydroxy butyrate, and polyhydroxybutyrate-hydroxyvalerate) and bacterial, and polyesters obtained from microorganisms (polylactic acid and polyglycolic acid). Synthetic biopolymers can be further classified into aliphatic polyesters and poly-vinyl-based esters. The cell structures of natural fibres are relatively complicated, and they are a composite of rigid cellulose microfibrils implanted in a soft lignin and hemicellulose matrix.

4.5 CONCLUSIONS

Sorption is a non-destructive process which is technologically simple (simple equipment) and adaptable to modifications. A wide range of materials can be used as sorbents for different types of pollutants. It is highly effective with very fast kinetics and produces a high quality of treated effluent. In order to make this process greener, the process of uptake by growing plants is being explored as an alternative green technique. This process is known as phytoremediation, which is discussed in the next chapter.

REFERENCES

1. E. Robens and S.A.A. Jayaweera, *Adsorp. Sci. Technol.* 32 (2014) 425–442.
2. A. Hauer, Sorption Theory for Thermal Energy Storage, In *Thermal Energy Storage for Sustainable Energy Consumption*, pp. 393–408, Springer, 2007.
3. V. Inglezakis and S. Poulopoulos, *Adsorption, Ion Exchange and Catalysis*, Vol. 3, pp. 498–520, Elsevier, 2006.
4. N.C. Srivastava and I.W. Eames, *App. Therm. Eng.* 18(9–10) (1998) 707–714.
5. M. Berhe Desta, *J. Thermody.* (2013) Article ID 375830, http://dx.doi.org/10.1155/2013/375830.
6. N. Ayawei, A. Ebelegi, and D. Wankasi, *J. Chem.* 2017 (2017) 1–11.
7. X. Yan, X. Fan, Q. Wang, and Y. Shen, *Therm. Sci.* 21 (2017) 1645–1649.
8. J.U. Keller and R. Staudt, *Gas Adsorption Equilibria: Experimental Methods and Adsorptive Isotherms*, pp. 359–413, Springer, 2005.
9. L. Kong and H. Adidharma, *Chem. Eng. J.* 375 (2019) 122112.
10. S. Azizian, S. Eris, and L.D. Wilson, *Chem. Phy.* 513 (2018) 99–104.
11. X. Chen, *Inform.* 6(1) (2015) 14–22.
12. D.H. Everett and E.A. Flood (Eds.), *In the Solid-Gas Interface*, Vol. 2, p. 1055, Marcel Dekker, 1967.
13. V.J. Inglezakis, S.G. Poulopoulos, and H. Kazemian, *Micropor. Mesopor. Mat.* 272 (2018) 166–176.

14. M. Sultan, T. Miyazaki, and S. Koyama, *Renew. Energy* 121 (2018) 441–450.
15. M. Khalfaoui, S. Knani, M.A. Hachicha, and A. Ben Lamine, *J. Coll. Interf. Sci.* 263 (2003) 350–356.
16. K.Y. Foo and B.H. Hameed, *Chem. Eng. J.* 156(1) (2010) 2–10.
17. Y.C. Wong, Y.S. Szeto, W.H. Cheung, and G. McKay, *Pro. Biochem.* 39 (2004) 695–704.
18. S. Hong, C. Wen, J. He, F. Gan, and Y.S. Ho, *J. Hazard Mater.* 167(2009) 630–633.
19. Y.-S. Ho, *Carbon.* 42 (2004) 2115–2116.
20. R. Han, Y. Wang, W. Zou, Y. Wang, and J. Shi, *J. Hazard. Mater.* 145 (2007) 331–335.
21. V. Kumar and S. Sivanesan, *Dyes and Pigm.* 72 (2007) 130–133.
22. Y.-S. Ho, *Water Res.* 40 (2006) 119–125.
23. D.M. Ruthven, *Principle of Adsorption and Adsorption Processes*, John Willey and Sons, 1984.
24. M.A. Al-Ghouti and D.A. Da'ana, *J. Hazard. Mater.* 393 (2020) 122383.
25. A.A. Inyinbor, F.A. Adekola, and G.A. Olatunji, *Wat. Resour. Ind.* 15 (2016) 14–27.
26. W.W. Hassler, *J. Am. Water Works Ass.* 33 (1941) 2124–2152.
27. H. Marsh and F. Rodríguez-Reinoso, *Activated Carbon*, Elsevier, 2006.
28. F. Rodríguez-Reinoso and M. Molina-Sabio, *Carbon* 30 (1992) 1111–1118.
29. C. Moreno-Castilla, F. Carrasco-Marín, M.V. López-Ramón, and M.A. Alvarez-Merino, *Carbon* 39 (2001) 1415–1420.
30. T. Yang and A.C. Lua, *J. Colloid Interf. Sci.* 267 (2003) 408–417.
31. C. Bouchelta, M.S. Medjram, O. Bertrand, and J.-P. Bellat, *J. Anal. Appl. Pyrolysis* 82 (2008) 70–77.
32. N. Baig, I. Kammakakam, and W. Falatha, *Mater. Adv.* 2 (2021) 1821–1871.
33. R. Shukla, D.P. Dutta, Jayshree Ramkumar, B.P. Mondal, and A.K. Tyagi, Nanocrystalline Functional Oxide Materials, In *Springer Handbook of Nanomaterials*, Ed. R. Vajtai, Springer, 2013.
34. Jayshree Ramkumar, *Arch. Nano Op. Acc. J.* 1(4) (2018), https://doi.org/10.32474/ANOAJ.2018.01.000116.
35. M.E.A. El-Sayed, *Sci. Tot. Environ.* 739 (2020) 139903.
36. Jayshree Ramkumar, R. Shukla, S. Chandramouleeswaran, T. Mukherjee, and A.K. Tyagi, *Nanosci. Nanotechnol. Lett.* 4 (2012) 693–700.
37. J. Mukherjee, Jayshree Ramkumar, S. Chandramouleeswaran, R. Shukla, A.K. Tyagi, *J. Radioanal. Nucl. Chem.* 297 (2013) 49–57.
38. J. Majeed, Jayshree Ramkumar, S. Chandramouleeswaran, and A.K. Tyagi, *Sep. Sci. Technol.* 50 (2015) 404–410.
39. J. Majeed, Jayshree Ramkumar, S. Chandramouleeswaran, and A.K. Tyagi, *Adv. Por. Mater.* 2 (2014) 1–10.
40. J. Majeed, Jayshree Ramkumar, S. Chandramouleeswaran, O.D. Jayakumar, and A.K. Tyagi, *RSC Adv.* 3 (2013) 3365–3373.
41. J. Majeed, Jayshree Ramkumar, S. Chandramouleeswaran, and A.K. Tyagi, *Sep. Sci. Technol.* 55 (2020) 1922–1931.
42. M. Bhaumik, T. Leswifi, A. Maity, V.V. Srinivasu, and M. Onyango, *J. Haz. Mater.* 186 (2011) 150–159.
43. U.O. Aigbe, R.B. Onyancha, K.E. Ukhurebor, and K.O. Obodo, *RSC Adv.* 10 (2020) 595–609.
44. M. Sujana and S. Anand, *App. Surf. Sci.* 256 (2010) 6956–6962.
45. M.G. Sujana, G. Soma, N. Vasumathi, and S. Anand, *J. Fluor. Chem.* 130 (2009) 749–754.
46 L. Chen, T.-J Wang, H.-X. Wu, Y. Jin, Y. Zhang, and X.-Min Dou, *Powd. Technol.* 206 (2011) 291–296.47. Y. Sun, F. Qinghua, D. Junping, C. Xiaowei, and X. Jiaqiang, *Desal.* 277 (2011) 121–127.
48. B.D. Turner, P. Binning, and S.L.S. Stipp, *Environ. Sci. Technol.* 39 (2005) 9561–9568.
49. M. Islam and R.K. Patel, *J. Hazard. Mater.* 143 (2007) 303–310.

50. S. Jain and R.V. Jayaram, *Sep. Sci. Technol.* 44 (2009) 1436–1451.

51. M. Mourabet, A.E. Rhilassi, H.E. Boujaady, M.B. Ziatni, R.E. Hamri, and A. Taitai, *Appl. Surf. Sci.* 258 (2012) 4402–4410.

52. T. Yang, C. Kim, J. Jho, and I.W. Kim, *Coll. Surf. A. Physicochem. Eng. Asp.* 401 (2012) 126–136.

53. N. Sakhare, S. Lunge, R. Rayalu, S. Bakardjiva, J. Subrt, S. Devotta, and N. Labhsetwar, *Chem. Eng. J.* 203 (2012) 406–414.

54. H. Liu, S. Deng, Z. Li, G. Yu, and J. Huang, *J. Haz. Mater.* 179 (2010) 424–430.

55. J. Kang, B. Li, J. Song, D. Li, J. Yang, W. Zhan, and D. Liu, *Chem. Eng. J.* 166 (2011) 765–771.

56. S. Sanuja, A. Agalya, and M.J. Umapathy, *Int. J. Polym. Mater.* 63 (2014) 733–740.

58. R. Shukla, Jayshree Ramkumar, and A.K. Tyagi, *Int. J. Nanotech.* 7 (2010) 989–1002.

59. Y.H. Li, S. Wang, X. Zhang, J. Wei, C. Xu, Z. Luan, D. Wu, and B. Wei, *Environ. Technol.* 24(3) (2003) 391–398.

60. A. Rajput, S.K. Raj, P.P. Sharma, V. Yadav, H. Sarvaia, H. Gupta, and V. Kulshrestha, *J. Disp. Sci. Technol.* 40 (2019) 1101–1109.

61. L.H. Velazquez-Jimenez, R.H. Hurt, J. Matos, and J.R. Rangel-Mendez, *Environ. Sci. Technol.* 48 (2014) 1166–1174.

62. S. Tripathy and A. Raichur, *J. Haz. Mater.* 153 (2008) 1043–1051.

63. S. Kumar, A. Gupta, and J.P. Yadav, *J. Environ. Biol.* 29 (2008) 227–232.

64. R. Kumar and J. Chawla, *Carbon-Based Materials for De-Fluoridation of Water: Current Status and Challenges, Carbon-Based Material for Environmental Protection and Remediation*, Ed. M. Bartoli, M. Frediani, and L. Rosi, IntechOpen, 2020, https://doi.org/10.5772/intechopen.90879.

65. S. Kagne, S. Jagtap, P. Dhawade, S.P. Kamble, S. Devotta, and S.S. Rayalu, *J. Haz. Mat.* 154 (2008) 88–95.

66. W.-H. Kang, E.-I. Kim, and J.-Y. Park, *Desal.* 202 (2007) 38–44.

67. S.V. Tarali, N.P. Hoolikantimath, N. Kulkarni, et al., *SN Appl. Sci.* 2 (2020) 1205.

68. S. George Suja, D. Mehta, and V.K. Saharan, *Rev. Chem. Eng.* 36 (2020) 369–400.

69. X. Yu, S. Tong, M. Ge, and J. Zuo, *Carb. Polym.* 92 (2013) 269–275.

70. Y. Nie, C. Hu, and C. Kong, *J. Haz. Mat.* 233–234 (2012) 194–199.

71. D.-W. Cho, B.-H. Jeon, Y. Jeong, I.-H. Nam, U.-K. Choi, R. Kumar, and H. Song, *App. Surf. Sci.* 372 (2016) 13–19.

72. T.L. Tan, P. Krushnamurthy, H. Nakajima, and S.A. Rashid, *RSC Adv.* 10 (2020) 18740–18752.

73. L.H. Velazquez-Jimenez, R.H. Hurt, J. Matos, and J.R. Rangel-Mendez, *Environ. Sci. Technol.* 48 (2014) 1166–1174.

74. M. Rajan and G. Alagumuthu, *J. Chem.* (2013) Article ID 235048, https://doi.org/10.1155/2013/235048.

75. S.K. Swain, S. Mishra, T. Patnaik, R.K. Patel, U. Jha, and R.K. Dey, *Chem. Eng. J.* 184 (2012) 72–81.

76. G. Zhang, Z. He, and W. Xu, *Chem. Eng. J.* 183 (2012) 315–324.

77. J. Cheng, X. Meng, C. Jing, and J. Hao, *J. Haz. Mater.* 278 (2014) 343–349.

78. S. Ayoob and A.K. Gupta, *Chem. Eng. J.* 150 (2009) 485–491.

79. S. George, P. Pandit, and A.B. Gupta, *Wat. Res.* 44 (2010) 3055–3064.

80. S. Ghorai and K.K. Pant, *Chem. Eng. J.* 98 (2004) 165–173.

81. A. Goswami and M.K. Purkait, *Chem. Eng. Res. Des.* 90 (2012) 2316–1324.

82. S.M. Maliyekkal, A.K. Sharma, and L. Philip, *Wat. Res.* 40 (2006) 3497–3506.

83. K. Selvi, S. Pattabhi, and K. Kadirvelu, *Biores. Technol.* 80 (2001) 87–89.

84. M.K. Rai, B.S. Giri, Y. Nath, H. Bajaj, S. Soni, R.P. Singh, R.S. Singh, and B.N. Rai, *J. Wat. Supp. Res. Technol. Aqua* 67(8) (1 December 2018) 724–737.

85. J. Zhang, T. Shang, X. Jin, J. Gao, and Q. Zhao, *RSC Adv.* 5 (2015) 784–790.

86. R. Labied, O. Benturki, A.Y. Eddine Hamitouche, and A. Donnot, *Ads. Sci. Technol.* 36 (2018) 1066–1099.

87. S. Parlayici, V. Eskizeybek, A. Avcı, et al., *J. Nanostruct. Chem.* 5 (2015) 255–263.

88. C. Namasivayam and M.V. Sureshkumar, *Biores. Technol.* 99 (2008) 2218–2225.

89. S.S. Baral, S.N. Das, and P. Rath, *Biochem. Eng. J.* 31 (2006) 216–222.

90. Z. Akmar Zakaria, M. Suratman, N. Mohammed, and W. Azlina Ahmad, *Desal.* 244 (2009) 109–121.

91. E. Elmolla, W. Hamdy, A. Kassem, and A. Hady, *Desal. Wat. Treat.* 57 (2015) 1–9.

92. B.V. Babu and S. Gupta, *Adsorption* 14 (2008) 85–92.

93. S.P. Mishra, *Curr. Sci.* 107 (2014) 601–612.

94. M.K. Mondal and R. Garg, *Environ. Sci. Poll. Res. Int.* 24(15) (2017) 13295–13306.

95. W. Chen, R. Parette, J. Zou, F.S. Cannon, and B.A. Dempsey, *Wat. Res.* 41 (2007) 1851–1858.

96. R. Sawana, Y. Somasundar, V.S. Iyer, et al., *Appl. Water Sci.* 7 (2017) 1223–1230.

97. Z. Shabbir, M. Shahid, K. Natasha, et al., *Environ. Geochem. Health* 45 (2020), https://doi.org/10.1007/s10653-020-00782-1.

98. S. Mukherjee, A. Kumar Thakur, R. Goswami, P. Mazumder, K. Taki, M. Vithanage, and M. Kumar, *J. Environ. Manag.* 281 (2021) 111814.

99. Y.N. Chen, L.Y. Chai, and Y.D. Shu, *J. Haz. Mat.* 160 (2008) 168–172.

100. A. Bhatnagar, V.J.P. Vilar, C.M.S. Botelho, and R.A.R. Boaventura, *Environ. Technol.* 32 (2011) 231–249.

101. H.S. Altundoğan, S. Altundoğan, F. Tümen, and M. Bildik, *Waste Manag. (New York, N.Y.)* 22(3) (2002) 357–363.

102. J. Plodinec, *Eur. J. Glass Sci. Technol. Part A* 41 (2000) 186–192.

103. S. Chandramouleeswaran, Jayshree Ramkumar, V. Sudarsan, and A.V.R. Reddy, *J. Haz. Mater.* 198 (2011) 159–164.

104. Jayshree Ramkumar and S. Chandramouleeswaran, *J. Chem. Appl. Biochem.* 4(1) (2017) 121.

105. Jayshree Ramkumar, S. Chandramouleeswaran, V. Sudarsan, R.K. Mishra, C.P. Kaushik, K. Raj, T. Mukherjee, and A.K. Tyagi, *J. Haz. Mater.* 154 (2008) 513–518.

106. S. Chandramouleeswaran and Jayshree Ramkumar, *J. Haz. Mater.* 280 (2014) 514–523.

107. Jayshree Ramkumar, and S. Chandramouleeswaran, *J. Radioanal. Nuc. Chem.* 298 (2013) 1543–1549.

108. Jayshree Ramkumar, S. Chandramouleeswaran, B.S. Naidu, and V. Sudarsan, *J. Radioanal. Nuc. Chem.* 298 (2013) 1845–1855.

109. J. Mukherjee, Jayshree Ramkumar, S. Chandramouleeswaran, R. Shukla, and A.K. Tyagi, *J. Radioanal. Nucl. Chem.* 297 (2013) 49–57.

110. A. Singha Deb, B. P. Mohanty, K. Ilaiyaraja, K. Sivasubramanian, and V. Balasubramaniam, *J. Radioanal. Nuc. Chem.* 295 (2012) 1161–1169.

111. Y.S. Ho and G. McKay, *Proc. Safety Environ. Protect.* 76 (1998) 183–191.

112. L.C. Morais, O.M. Freitas, E.P. Goncalves, L.T. Vasconcelos, and C.G. Beca, *Wat. Res.* 33 (1999) 979–988.

113. S. Venkata Mohan, N. Rao, and J. Karthikeyan, *J. Haz. Mater.* 90 (2002) 189–204.

114. Y.S. Ho and G. McKay, *Chem. Eng. J.* 70 (1998) 115–124.

115. L. Sepúlveda, K. Fernández, E. Contreras, and C. Palma, *Environ. Technol.* 25 (2004) 987–996.

116. U. Filipkowska, *Adsorpt. Sci. Technol.* 24 (2006) 781–795.

117. D.C. da Silva Alves, B. Healy, L.A.D.A. Pinto, T.R.S. Cadaval, Jr., and C.B. Breslin, *Mol.* 26 (2021) 594.

118. A.A. Adeyemo, I.O. Adeoye, and O.S. Bello, *Appl. Water Sci.* 7 (2017) 543–568.

119. K.S. Bharathi and S.T. Ramesh, *Appl. Water. Sci.* 3 (2013) 773–790.

120. A.K. Jain, V.K. Gupta, A. Bhatnagar, and Suhas, *Sep. Sci. Technol.* 38 (2003) 463–481.

121. Jayshree Ramkumar, S. Chandramouleeswaran, V. Sudarsan, R. Vatsa, S. Shobha, V.K. Shrikhande, G. Kothiyal, and T. Mukherjee, *J. Non-Cryst. Solids* 356 (2010) 2813–2819.
122. Jayshree Ramkumar, S. Chandramouleeswaran, V. Sudarsan, R. K. Mishra, C.P. Kaushik, K. Raj, and A.K. Tyagi, *J. Hazard Mater* 172 (2009) 457–464.
123. D.P. Dutta, A. Mathur, Jayshree Ramkumar, and A.K. Tyagi, *RSC Adv.* 4 (2014) 37027–37035.
124. D.P. Dutta, A. Singh, Jayshree Ramkumar, K. Bhattacharyya, A. Tyagi, and M. Fulekar, *Adv. Por. Mater.* 2 (2014) 237–245.
125. A. Singh, D.P. Dutta, Jayshree Ramkumar, K. Bhattacharya, A.K. Tyagi, and M.H. Fulekar, *RSC Adv.* 3 (2013) 22580–22590.
126. J. Mukhopadhyay, D.P. Dutta, Jayshree Ramkumar, and A.K. Tyagi, *J. Environ. Chem. Eng.* 4 (2016), https://doi.org/10.1016/j.jece.2016.06.015.
127. Jayshree Ramkumar, J. Hisham Zain, and S. Chandramouleeswaran, *I. Micropor. Mesopor. Mater.* 314 (2021) 110858, https://doi.org/10.1016/j.micromeso.2020.110858.
128. Jayshree Ramkumar and S. Chandramouleeswaran, *Chapter 30 in Green Polymer Composite Technology: Properties and Applications*, Ed. Inammudin, CRC Press, 2016.

5 Phytoremediation
An emerging green auxiliary technique

5.1 INTRODUCTION

Bioremediation is an important process for the removal of pollutants using living or dead organisms [1]. When living organisms utilize pollutants as their source of food, they break them in situ or ex situ into carbon dioxide and water at optimal temperature. The process is slow, but being a natural process, it is benign, less labour-intensive, and cost-effective. Some of the bioremediation methods are bioaugmentation, rhizofiltration, biostimulation, phytoremediation, mycoremediation, composting, etc. In this chapter, the focus will be on phytoremediation of pollutants. Phytoremediation is a subset of bioremediation wherein living plants and vegetation are used to destroy toxic pollutants present in environmental media (surface water, ground water, waste water, sediments, soils, and/or external atmosphere). The word "phytoremediation" is derived from the Greek word "phyton", meaning "plant", and the Latin word "remedium", which means "to remedy/restore". Phytoremediation can be considered as an in situ method of bioremediation possessing the intrinsic benefits of bioremediation [2]. The removal of toxic species from ground water and soil by plants can occur through different mechanisms. The interface of plants with the environmental medium and microorganisms present plays a very crucial role. The competence of the process is largely dependent on the nature of the contaminant, plant species, and medium.

5.2 TYPES OF PHYTOREMEDIATION

The classification of the different phytoremediation strategies is based upon the characteristics of contaminants and plants. Figure 5.1 shows the schematic representation of the different processes (drawn based on Ref. 3).

(i) Phytostimulation (biodegradation of enhanced rhizosphere or rhizosphere degradation or "rhizo degradation") is a process in which the microbial activity of soil (surrounding the roots of a plant) is enhanced to degrade contaminants by organisms of the roots by aerobic conversions (through oxygenate rhizosphere) and also increase the bioavailability of organic carbon. The distribution of nutrients is heterogeneous, thus leading to varying spread of microbes. This microbial degradation in the rhizosphere of media such as soil sediment and sludge is stimulated by plants like grass, alfalfa, *Cassia*, and *Typha* for the degradation of organic compounds like herbicides, phenols, and chlorinated solvents.

DOI: 10.1201/9781003442516-5

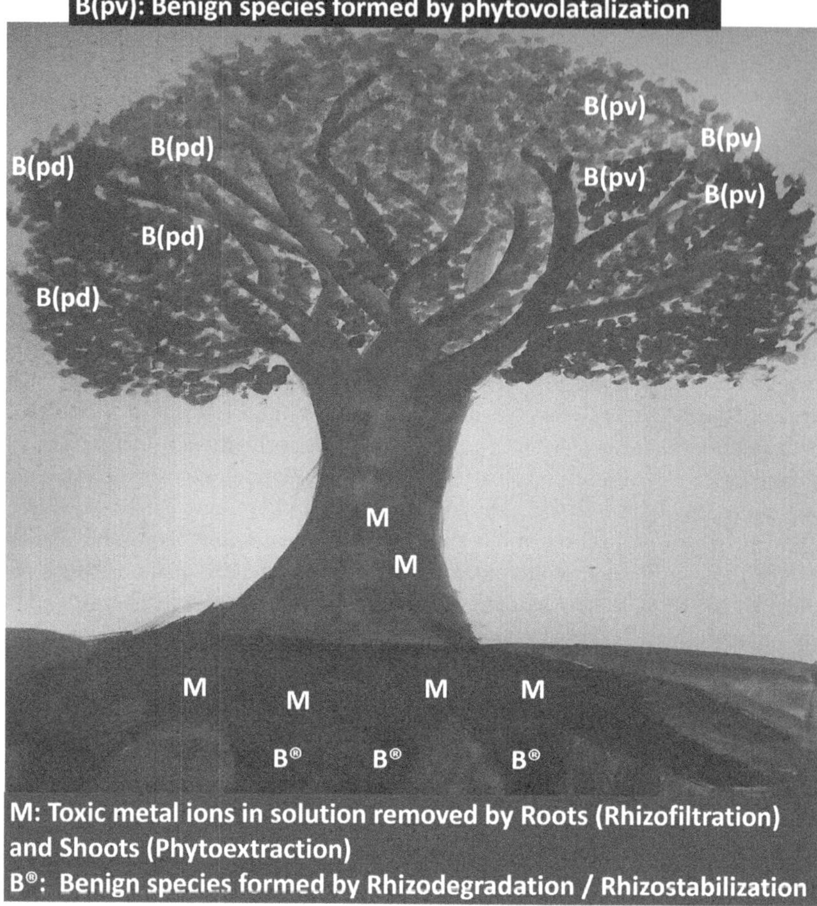

FIGURE 5.1	Schematic representation of different phytoremediation processes

(ii) Phytoextraction is the uptake (phytoabsorption), accumulation (phyto-accumulation), and concentration of pollutants by the roots of plants and amassing into their aerial parts (shoots/leaves). Such plants are known as hyperaccumulator plants (*Pteris vittata*, *Elsholtzia splendens*, *Thlaspicaerulescens*, and *Alyssum bertolonii*), and they can concentrate specific pollutants to as high as 1% dry weight depending on the nature of the pollutant. The specific species can then be retrieved from these parts of the plants. This is also known as phytoaccumulation and involves the sorption of toxic metals by plant roots followed by translocation of sorbed metals to shoots and deposition at the vacuole, cell wall, cell membrane, and other metabolically inactive parts in plant tissues. Hyperaccumulator plants accumulate a higher concentration of toxic metals in their root and shoot tissues by the formation of a metal–phytochelatin complex (M-PC) or metal–ligand complex formation inside plant cells, which are translocated to the plants' vacuole in which the ions are amassed. The removal efficiency of a plant is controlled by the plant biomass, and so the plant should grow fast to be suitable for the process. This involves the extraction of metal ions like Co, Cu, Ni, Pb, Zn, Hg, Mo, Ag, Cd, Sr, Cs, and U from soil sediments and sludge within the shoots of plants like sunflower and *Brassica*.

(iii) Phytodegradation (phytotransformation) is the enzyme-catalysed [laccases (for anilines), dehalogenases (for chlorinated solvents and pesticides), and nitroreductases (for nitro-aromatic compounds)] breakdown of organic contaminant in tissues of plants (*Myriophyllum spicatum*, *Canna*, and *Populus*). The changes caused results in degradation, inactivation, or immobilization (phytostabilization), or conversion to a non-toxic form. Thus, the plant can lead to metabolism and subsequent destruction of toxic species within plant tissues. This is more appropriate for organic pollutants. It is to be understood that the nature of the plants plays a critical function directly or indirectly in this process. The direct process involves the absorption of pollutants by the plants and break down into non-toxic compounds inside the plant tissue. However, if the pollutants are degraded within the soil and the plants tend to absorb the benign by-products, the process is known as rhizodegradation, and the plants have an indirect role in the process.

(iv) Phytostabilization (phytoimmobilization) is the process by which toxic species are immobilized within the rhizosphere or roots of plants. The mechanism involves the formation of chelates by the chelating agents present within certain plants like *Alyssum*, *Haumaniastrum*, *Eragrostis*, *Gladiolus*, and *Ascolepis*. Thus, chelation tends to immobilize metal pollutants into the rhizosphere or roots. The fibrous roots of phreatophytic trees could be used for controlling water flow and soil erosion in addition to the uptake of heavy metal ions $(Cu, Pb, Zn, Cu, As, U, Se)$. Rhizostabilization involves the fixing of metal ions (As, Cd, Cu, Cr, Pb, Zn) within the soil sediment by complexation using the roots of plants like sunflower and *Chenopodium*.

(v) Phytovolatilization is the ability of certain plants to convert some of the absorbed metals/metalloids (Hg, Se, As, and elements IIB, VA, and VIA of the periodic table) by the roots into a benign form and then release it into the atmosphere. The contaminants are translocated to the leaves and are volatilized through plant stomata (with sites for gaseous exchange for plant transpiration). This involves many sequential stages. The main requirement of this technique is that volatile products should be less toxic than heavy metal ions. The process of volatilization can be direct or be influenced by the interactions between the root and the soil. The discharge can occur through hydrophobic barriers like cuts, epidermis, suberin, and other dermal layers. At times, there is an upward movement, and release occurs during transpiration. This process is used to remove chlorinated solvents, Hg, Se, and As from the soil sediment using plants like *Populus* and *Scirpus*.

(vi) Rhizofiltration (phytofiltration) is the process through which plants (*Fontinalis antipyretica, Helianthus annuus, Phragmites australis, Brassica juncea*, and several species of *Callitriche, Salix, Lemna*, and *Populus*) concentrate, sorb, and/or precipitate pollutants through the plant root system. The contaminants remain sorbed in the roots, which are specifically grown to a large size prior to their contact with contaminated water. Once the roots are saturated, the plants are harvested and disposed of safely. This process can be used to remove ions like Cd, Cu, Ni, Zn, and Cr from water bodies using roots of plants like *Lemna*.

Of all the techniques, phytoextraction and phytostabilization are most useful [4]. Plants used for phytoextraction should be easy to farm, grow rapidly, and result in high biomass for the process to be economical. In addition to this, the plants should have a copious root system to absorb the metal ions and excellent tolerance to high metal concentrations [5]. Phytoextraction is effective only if the accumulated contaminant is subsequently removed through harvesting before leaf fall or death, warranting that contaminants do not diffuse back to the soil [6]. The processing of biomass to obtain metals is known as phytomining [2]. Alternatively, the biomass can be incinerated. The commercial value recovery of precious metals (Ni, Zn, Cu, and Co) and energy production by incineration [7] are factors that favour the phytoremediation process. These processes need great care to avoid further contamination or pollution.

The efficiency of the different processes depends on the optimization of the two key parameters, namely, the ability to hyperaccumulate metals and the yield of biomass. The affixing of organic or inorganic reagents to contaminated soil improves extraction by the alteration of metal speciation and reduction of heavy metal solubility and bioavailability by changing the pH value and redox status of the soil [8–12]. The increase in organic matter content and essential nutrients in soil leads to enhanced performance of the plant. The mechanism involves the assistance of microorganisms in the rhizosphere (bacteria and mycorrhiza) for phytostabilization, resulting in improved metal immobilization due to sorption on cell walls which contain chelators [13, 14]. Cell walls also behave as a filtration barrier to the translocation of metal ions from roots to shoots [15]. The different parameters that affect the extraction efficiency are plant type, heavy metal bioavailability, soil, and rhizosphere properties. Appropriate selection of the plant species is vital for effective phytoextraction [16, 17]. Therefore, two different strategies are adopted for the selection of the plant, which involves the use of hyperaccumulator plants (which can accumulate heavy metals in above-ground parts to a greater extent) or plants with high above-ground biomass production (even with lower metal-accumulating capacities) [18, 19]. Plants which accumulate high concentrations of heavy metals in their aerial parts without showing any phytotoxicity symptoms are known as hyperaccumulators [20, 21]. It has been observed that hyperaccumulators are found to accumulate nearly 100-fold greater concentration levels as compared to normal plants under the same conditions [20]. Thus, for a plant to be defined as a hyperaccumulator, the ratio of concentrations of heavy metals in the shoot with respect to that in soil and the root is greater than 1. This indicates effective uptake from soil and transport through the plant, respectively

[22, 23]. Moreover, the concentration levels of metal ions in the shoot should be of greater than a specified value (10 mg/kg for Hg; 100 mg/kg for Cd and selenium (Se); 10,000 mg/kg for Co, Cu, Cr, Ni, and Pb; and 10,000 mg/kg for Zn and Mn) [24]. Thus, choosing the right hyperaccumulator is critical for the successful phytoremediation of heavy metals. More than 450 plant species (Brassicaceae, Fabaceae, Euphorbiaceae, Asteraceae, Lamiaceae, and Scrophulariaceae families) [19, 25] have been recognized as hyperaccumulators [26]. It is reported that some species can even accumulate more than two elements, e.g. *Sedum alfredii* can hyperaccumulate Zn, Pb, and Cd [28-30]. It is important to ensure that edible plants are not used for phytoextraction to avoid introduction of heavy metals into the food chain. The use of plant growth-promoting rhizobacteria (PGPR) has gained popularity due to the fact that PGPR promotes plant growth, protects plants from pathogens, and enhances the metal tolerance and uptake property [14, 30, 31]. Arbuscular mycorrhizal fungi (AMF) is an important microbe used to assist phytoremediation by increasing the surface area of roots (network formation) [32] and also the production of phytohormones [33].

Bioconcentration factor (or biological absorption coefficient) and translocation factor are the two parameters which indicate phytoextraction efficiency. Bioconcentration factor $\left(\mathrm{BCF} = \mathrm{C}_{\mathrm{plant}} / \mathrm{C}_{\mathrm{soil}} \right)$ is the ratio of the total concentration of an element in the harvested plant tissue $\left(\mathrm{C}_{\mathrm{plant}} \right)$ to its concentration in the soil in which the plant was growing $\left(\mathrm{C}_{\mathrm{soil}} \right)$ [34, 35]. Translocation factor $\left(\mathrm{TF} = \mathrm{C}_{\mathrm{plant}} / \mathrm{C}_{\mathrm{root}} \right)$ is the ratio of the total concentration of elements in the aerial parts of the plant $(\mathrm{C}_{\mathrm{shoot}})$ to the concentration in the root $\left(\mathrm{C}_{\mathrm{root}} \right)$ [36]. The commercial efficiency of phytoextraction is estimated as the amount of metal ions removed per hectare in a year. This is the mathematical product of rate of metal accumulation [metal (g)/plant tissue (kg)] and biomass yield [plant tissue (kg)/hectare/year]. This value should be at least 1 kg/ha/year for it to be commercially viable, but even then, phytoremediation may occur in 15–20 years. Hence, it is not practical to wait for an unrealistic time period, and fast growing crops are used instead [37].

There are other phytoremediation strategies which are combinations or variations of the above mechanisms. These are discussed as follows:

(i) Hydraulic barriers: Deep-rooted large trees like *Populus* sp. consume large quantities of groundwater during transpiration. The contaminants present in the water are metabolized by plant enzymes and vaporized together with water or simply sequestered within the plant tissues [38].

(ii) Vegetation covers: Grasses and trees on landfills or tailing behave as a vegetation cover, minimizing rain water infiltration and averting the dispersal of pollutants. Soil aeration is increased, resulting in biodegradation, evaporation, and transpiration. The complexity of this technique is that tailings are not suited for root development [39].

(iii) Constructed wetlands consist of organic soils, microorganisms, algae, and vascular aquatic plants in areas where the water level is at/near the surface. However, the technique is mainly used in the treatment of domestic water, agricultural water, industrial waste water, and acid mine drainages due to its advantages like high removal efficiency, low assembly rate, effortless manoeuvre, and maintenance [40].

(iv) Phytodesalination is a technique using halophytes to remove excess salts (from saline soils), and this is similar to phytoextraction [41].

5.3 STEPS INVOLVED IN THE PROCESS

The general process of phytoremediation consists of various steps which are usually followed in a linear manner. The first step is plant selection and positioning. Since the process of phytoremediation is time consuming, it is very essential to make the correct decision on the type and quantity of the plants required. The nature of the soil usually determines the type of plant that is appropriate. However, extensive studies are needed on the effects of toxins on the plants. Usually plants which yield large amounts of biomass (more than 3 tonnes/acre) are chosen. Plants known as "hyperaccumulators" are very much needed for circumventing inorganic pollutants, as these plants will take up very high levels of these pollutants as compared to the soil itself. Deep-rooted trees like poplar, cottonwood, and willow are normally planted in rows, perpendicular to the flowing groundwater, to remove pollutants. The next step is irrigation and soil amendment. This is important because, more than often, solids need be altered and irrigated to ensure plant viability. At times, flooding of soil is carried out to enhance pollutant dissolution. Removal of pollutants can be accompanied with the depletion of nutrients from the soil. Therefore, soil pH is affected and the use of chelating agents can retain the metal ions in the solution, thereby enhancing the chances of phytoextraction. Monitoring is the next step, which has to be carried out on a regular schedule. The effectiveness of the process is measured from the change in the contaminant concentration over a time period. If the process is too time-consuming, an alternate protocol involving either using a different plant species or soil amendments can be carried out. Harvesting is an essential step which involves the removal of plants including the roots (containing contaminants). Therefore, a simple and efficient disposal protocol is needed. One of the most attractive procedure is the process of incineration, which results in the oxidation or vaporization of contaminants. There are instances when the biomass with the contaminant can be used as such. This is usually seen in the case of Se, which is now recognized as an essential trace element for ruminants. It is required in cattle for normal growth and fertility and for helping prevent other health disorders such as mastitis and calf scours. Therefore, the biomass with Se can be sold to livestock farmers as bioenhanced and reduce the cost of purchasing Se salts to be added to normal livestock feed [42]. Similarly, ash from the nickel-accumulator *Alyssum murale* can be economically processed into ore if the content is above 20% by weight [42]. It is usually seen that inorganic pollutants are mostly retained in the roots followed by the leaves and stems, with the least retention in the fruits. However, for nightshade plants like cucumbers and tomatoes, it has been observed that the concentration is much higher in their leaves and stems than in the roots.

Phytomining is the process of recovery of extracted heavy metals from biomass [43]. The plant biomass with the accumulated metals is then burnt in oxygen to obtain energy, and the remaining ash, known as bio-ore, is treated for recovery of the heavy metal. Thus, phytomining offers a dual advantage of production of energy and recovery of precious metals from biowaste. Studies have showed that the energy obtained is quite comparable to that from other sources and also leads to reduction of air pollution to a great extent. It has also been observed that processing of bio-ores results in low SO_X emissions into the atmosphere due to the low sulphur content, making phytomining an environmentally benign and cost-effective process. But the main controlling factor that makes this process commercially viable is its phytoextraction efficiency and the market for the recovered heavy metal ions. Phytomining is being carried out commercially for nickel, and the method has been found to be more attractive than conventional extraction methods.

Another aspect of removal of heavy metals is the role of plant phytochelatins (PCs) and metallothioneins (MTs) in phytoextraction [44]. These are loaded with cysteine sulphadryls which bind and sequester heavy metal ions in very stable complexes. PCs are small glutathione-derived enzymatically synthesized peptides, which bind metals and are a principal part of the metal detoxification system in plants.

5.4 MECHANISMS

The removal of metal ions by phytoextraction from environmental media like water is dependent on factors like pH, interfering ions, biomass, organic matter, and exudates. Although heavy metal ions degrade soil and water resources, posing serious threats to human life, it is often seen that plants can grow in contaminated soil and water bodies without any undesirable consequences. Some plants eliminate the metals due to decreased uptake, while some plants show tolerance to high concentration levels due to complexation [45, 46]. Heavy metal tolerance is the ability of plants to survive in soil which is toxic to other plants [47]. When a particular metal is accumulated by the roots, it may be taken to root tissues (phytoimmobilization) or may be translocated to the aerial shoot parts of the plant through xylem vessels by symplastic and/or apoplastic pathway and accumulated in vacuoles (cellular organelles with low metabolic activity) [48]. This is done to avoid the interference of detrimental metals with important cellular metabolic processes. In order for a plant to be useful for phytoremediation, it needs to have high tolerance of trace metals as accumulation within plant tissues occurs. This characteristic of a plant is dependent on active metal ion transport into the vacuoles, cell wall metal binding, and complex formation (with proteins and peptides).

The corporeal task of ion channels was known for a very long time. The membranes of nerve and muscle cells show variations in electrical membrane potential depending on the surrounding milieu [49]. It was later discovered that the currents for Na^+ and K^+ are independent of each other. This could be attributed to the fact that the membranes contain separate and specific ion channels for Na^+ and K^+ conduction. These ion channels were made of specific transmembrane proteins [50–54]. Ion channels do not require an energy source to function, and ions flow only if there is imbalance inside and outside of a cell. Thus, electrochemical gradient (combination of electrical transmembrane potential and ion concentration gradient across a membrane) leads to the passage of ions through an ion channel [51]. The work on ion channels was so important that the Nobel Prize in Chemistry 2003 was awarded "for discoveries concerning channels in cell membranes" jointly with one half to Peter Agre "for the discovery of water channels" and with one half to Roderick MacKinnon "for structural and mechanistic studies of ion channels". There are four indistinguishable symmetrically arranged subunits, two outer (M1) and two inner (M2) transmembrane α-helices [53], that surround the ion conduction pore. The channel has regions of different diameters. The tapered region contains the K^+-specific protein, resulting in the functioning of a potassium-selective filter. The structure contains five oxygen atoms (of the carbonyl group) per subunit in the direction of the ion conduction pathway, and K^+ ion is located at the centre of a caged structure with 8 oxygen atoms. Potassium ion conduction occurs when there is a complexation or an alteration of membrane potential. The process of alternate conduction/non-conduction, known as gating, occurs due to conformational changes in protein leading to the occlusion or abetment of the ionic conduction pathway [54]. When the ionic channel is closed, the pore is obstructed at the intracellular gate by a bundle of inner transmembrane helices (M2). The ion channel measures 12 Å in the centre and tapers down to a narrow region with a diameter of 3.5 Å, wherein the inner helix bundle (M2) comes together at an angle tilted away from the pore axis [55]. The intracellular gate of the channel is located right below the cavity. At this point, the electrostatic repulsion between any two K^+ ions in adjacent sites is very large (40 kcal/mol), indicating the non-occupancy of the neighbouring sites due to unfavourable energy conditions. The conductance is very high (allowing more than 1,000 ions to pass per millisecond) [56]. Therefore, it is clear that pore diameter, solvation, and binding play a major role in selective transport. The literature reveals that KcsA shows two different pore conformations, enhancing potassium ion conduction [57].

Phytoremediation is a summation of chronological processes like metal fixing, uptake by roots, transport from the roots to shoots, incorporation of metal ions into the xylem, cellular compartmentalization, and sequestration [58]. In general, heavy metal ions are present as insoluble precipitates in soil, making it arduous for the plants to extract the ions. Therefore, root exudates are released by the plants to alter the pH conditions of the rhizosphere, facilitating the uptake of metal ions by the roots [59] by apoplastic (passive diffusion) and symplastic (active transport against electrochemical potential gradients and concentration across the plasma membrane) pathways [60]. The heavy metal ions form complexes with various chelating agents after entering into the root cells [17] or are sequestered inside the vacuoles, enter into the xylem, and are subsequently translocated to the shoots [48] and leaves [61] by specialized transporters [62, 63]. The members of the *ZIP* gene family present in the roots of *Thlaspicaerulescens* and *Arabidopsis halleri* prove to be excellent metal transporters (e.g. cadmium, iron, manganese, and zinc) [64]. Other proteins like HMA3, a vacuolar P1B-ATPase, are excellent bioaccumulators (tend to sorb/accumulate heavy metal ions, which might perhaps be functional for their development) of heavy metal ions [65–68], with an extremely long-distance root-to-shoot translocation of Zn and Cd [68]. The metal transporters involved in the regulation of metal homeostasis lead to the translocation of metals like Zn and Ni towards the internal partition [69]. It is reported that MTP1, an antiporter located on the vacuoles and plasma membrane, can lead to the bioaccumulation of zinc [70] and also of nickel in the vacuoles [71]. Natural resistance-associated macrophage proteins (NRAMPs) transport heavy metal ions including copper and cobalt [71]. Apart from these metal transporter proteins, complexing agents like citrate and histidine can result in the chelation of metal ions like iron and nickel [72, 73]. The main requirement of phytoremediation is that detoxification has to be carried out. This is done using two different strategies, namely, avoidance and tolerance [74]. Avoidance is the capability of plants to demonstrate diminished uptake of heavy metal ions and additionally curb their transport within the plants [59]. This is achieved due to reduced uptake by roots and metal ion precipitation by complexation with chelators [59], thus restraining toxicity at the extracellular level. A similar barrier reduces the transport from the root to the shoot system of plants, thus preventing the aerial parts from these heavy metal ions [74]. The incorporation of metal ions within plant cell walls is also an avoidance mechanism [75]. Cell wall pectin contains carboxylic acid groups which can form complexes and also behave as ion exchangers [76–79]. The chelation can be caused by the presence of different organic ligands in various parts of plants [80–83]. The metal chelates are transported from the cytosol and stored into inactive spaces like vacuoles [84–86], thus decreasing their toxic effects. However, if the amount of metal ions in the plants is of higher concentration levels and cannot be mitigated by the above methods, there is an enhanced production of enzymatic and non-enzymatic antioxidants [87–89]. This has been found to be an excellent alternative mitigation process adopted by many plants.

5.5 NANO-PHYTOREMEDIATION

This technique is a combination of nanotechnology and phytoremediation. Due to the large surface area and small size, nanomaterials can be utilized to enhance phytoremediation efficiency. The large surface area contributes to increased surface activity, while the small particle size makes it easy to access contaminated sites [90]. The efficiency of this process is affected by the properties of contaminants, plants, and nanomaterials as well as the environmental conditions. The characteristics of the contaminants that play a major role are molecular weight, type of bonds, mobility, flexibility, toxicity, hydrophobic and hydrophilic interactions, and reactivity. Properties of a plant such as its type, degree of branching, nature of the root system, tolerance capacity and accumulation ability, fast growth, and high biomass are crucial for the right choice of phytoremediation plants. The structure, size, shape, composition, concentration, and surface properties of nanomaterials play a crucial role in the selection of an appropriate nanomaterial. Apart from these, environmental conditions such as temperature, pressure, pH, and humidity and also the nature of minerals and microorganisms present around the plant have a definite impact on phytoremediation efficiency. The interactions between plants and nanoparticles are affected by the size and penetration ability of the nanomaterials [91]. Once the nanomaterials enter the plant roots, there are two modes of transport within the plant tissues, namely, apoplastic transport (occurs outside the plasma membrane, xylem vessels) [92] and symplastic transport (movement of water between the cytoplasm and sieve plates) [93]. The different applications have been discussed in the Applications section of the chapter. The application of nanomaterials to the soil for the remediation process is site-specific and is carried out using methods such as injection under pressure, recirculation, pressure pulse technology, and hydraulics. Some of the challenges of nano-phytoremediation need to be overcome and are as follows: (i) this is a new field, hence studies are few; (ii) more realistic studies need to be performed; (iii) long-term studies are essential to assess the impact of nanomaterials on soil characteristics and their toxicity; and (iv) studies focussing on ways to avoid the aggregation of nanomaterials and (v) on understanding the effect of meteorological conditions need to be conducted.

5.6 BENEFITS AND CONSTRAINTS OF PHYTOREMEDIATION

Like any process, phytoremediation has both advantages and disadvantages [94]. The low cost of the process is overshadowed by the long time period of study. Moreover, the nature and concentration of pollutants, the correct choice of plants, and the possible side effects have to be considered in detail prior to opting for this technique. The process of phytoremediation has several advantages. The process is in situ and passive in nature and is low cost as it uses solar energy. This has a reduced impact on the environment. The presence of plants leads to the reduction of dispersion of contaminants by wind and also by surface runoff and leaching. Harvesting is a well-established technique for the extraction of accumulated metals, and this technique is quite viable. Despite several advantages, phytoremediation is still in the infancy stage, and not much information is available regarding the remediation properties of different plants. The process is limited to surface and shallow soils and is very slow. The metal ions taken up can prove to be toxic to the food chain if animals eat plants with large quantities of heavy metal ions. The process is not adapted to climatic and environmental conditions. The compounds from plants may percolate into groundwater, and this becomes difficult to handle if the nature of degraded products is hazardous.

5.7 APPLICATIONS

A number of plants have been used for their applications in the field of phytoremediation [95–105]. Despite the availability of many plants as hyperaccumulators, their short life, poor biomass yield, and slow growth rate prove to be a hindrance to the process. So the possible use of plants that produce high biomass but are normally not known to be hyperaccumulators could be evaluated for their phytoextraction efficiency. In such instances, it has been observed that although these plants assimilate lower levels of heavy metals in their above-ground tissues, the high biomass production can circumvent this limitation and enhance the overall absorption compared to hyperaccumulators [106–108]. Thus, crops like *Helianthus annuus*, *Cannabis sativa*, *Nicotiana tabacum*, and *Zea mays*, which produce high biomass, have been useful for the removal of heavy metal ions [109–111]. Due to several hindering factors, large-scale applications of hyperaccumulators are limited [112]. The limitation can be overcome using protocols such as plant hybridization or genetic engineering to generate transgenic plants [89, 113]. Chemical mutagen [ethyl methane sulphonate (EMS)]-treated sunflowers produced giant mutant sunflowers with very high phytoextraction efficiency [114]. Genetic engineering also shows potential for improving phytoextraction efficiency. The genetic modification is carried out by the transfer and insertion of a gene from a foreign source (plant, bacteria, or animals) into the target plant. This process is more advantageous as it allows the modification of plants with required traits within a shorter time period [115]. The selection of genes for genetic engineering is very crucial to achieve the desired results. It has also been observed that heavy metals produce ROS, leading to oxidative stress. Therefore, it is important to enhance antioxidant activity to improve tolerance to metal ions [116]. Genetic engineering involved gene introduction, which could facilitate the processes of uptake, translocation, and sequestration [117–119]. Despite its advantages, genetic engineering has some limitations. Since heavy metal accumulation is a complex process, manipulation of multiple genes is essential, and this is laborious and time-consuming. At times, the modified plant may also prove to be toxic in nature. Thus, the use of microorganisms to stimulate root proliferation and to subsequently lead to improved plant growth with increased heavy metal tolerance is gaining popularity [120]. Plant growth-promoting rhizobacteria (PGPR) produce various compounds and follow different paths to enhance growth [14, 33, 121, 122]. The use of nano-phytoremediation for the removal of heavy metal ions is an attractive technique [123]. Nano-zero-valent iron (NZVI) is used extensively as an effective sorbent, reductant, and catalyst for many heavy metal ions [124–130]. The use of waste water (industrial, municipal, and household liquid waste) for irrigation and decontamination has also been evaluated [131].

It is of great interest to explore the plausibility of phytoremediation for treating dye-contaminated water bodies. As the study is in the initial stages, it is not very clear about the mechanism involved. The use of phytoremediation for dye removal is a research in a new dimension due to the structural complexity and chemical characteristics of the dyes. It is further to be understood that the presence of the dyes not only affects the aesthetic value of water but also alters certain parameters, which could hamper the phytoremediation process. The studies have been carried out using different types of weeds, ferns, grasses, or agricultural wastes. *Phragmites australis* could be used for the removal of textile dyes like acid orange 7 [132], while *Lemna minor* L. was found to be efficient for many dyes [133, 134]. Different mechanisms of treatment of waste water using plants like degradation, accumulation, dissipation, and immobilization are reported. The tolerance of *Lemna minor* plants and its stress response were evaluated using two triphenylmethane dyes (crystal violet and malachite green) [135]. It was seen that both phytoextraction and phytodegradation processes played a major role in the treatment of the dye-contaminated water stream. The biodegradation of brilliant green dye using callus cultures of *Tecoma stans* var. *angustata* plants [136] showed that the dye was completely degraded to minor non-toxic metabolites via complete degradation of the aromatic rings and by cleavage of functional groups. The removal of methylene blue (MB) and methyl orange (MO) has been achieved using the roots of water hyacinth *Eichhornia crassipes* [137]. The decolourization of malachite green could be achieved using *Blumeamalcolmii* due to the involvement of laccase, veratryl alcohol oxidase, and dichlorophenol indophenol reductase enzymes [138]. It is also seen that the presence of plant microbiome, plant-associated microorganisms (fungi/bacteria), enhances the decolourization process due to their ability to transform organic and inorganic compounds. It has been observed that *Medicago sativa* L. and *Sesbania cannabina* Pers. plants could be used in conjunction with *Gracilibacillus* sp. GTY (a salt-tolerant bacteria) to easily degrade acid red B or acid scarlet GR dyes [139]. Nano-phytoremediation has been used for the removal of dyes. However, this area of research deals with the use of nanomaterials synthesized using plant extracts.

5.8 CONCLUSIONS

The uptake of toxic species by plants is known as phytoremediation. This technology is gaining importance due to its unique advantages. However, since the technique is still in baby steps, not much work has been reported. Therefore, no clearcut strategy can be obtained. However, this provides a large scope for future work in this expanding field. It has been observed that basic chemical processes such as precipitation occur during the phytoremediation process. Thus, it can be understood that precipitation forms the basis for the separation of metal ions. In the next chapter, the uses of processes such as precipitation and flocculation with respect to metal ion separation are discussed.

REFERENCES

1. I. Sharma, *Bioremediation Techniques for Polluted Environment: Concept, Advantages, Limitations, and Prospects, Trace Metals in the Environment – New Approaches and Recent Advances*, Ed. M. Alfonso Murillo-Tovar, H. Saldarriaga-Noreña, and A. Saeid, IntechOpen, 7 December 2020, https://doi.org/10.5772/intechopen.90453, https://www.intechopen.com/chapters/70661.
2. V.C. Pandey and O. Bajpai, Chapter 1 – Phytoremediation: From Theory Toward Practice, In *Phytomanagement of Polluted Sites*, Ed. V.C. Pandey and K. Bauddh, pp. 1–49, Elsevier, 2019.
3. A. Yan, Y. Wang, S.N. Tan, M.L. Mohd Yusof, S. Ghosh, and Z. Chen, *Front. Plant Sci.* 11 (2020) 359, https://doi.org/10.3389/fpls.2020.00359.
4. B. Lorestani, N. Yousefi, M. Cheraghi, and A. Farmany, *Environ. Monitor. Asses.* 185 (2013) 10217–10223.
5. H. Ali, E. Khan, and M.A. Sajad, *Chemosphere* 91(7) (2013) 869–881.
6. G. Pierzynski, P. Kulakow, L. Erickson, and L. Jackson, *J. Nat. Res. Life Sci. Ed.* 31 (2002) 31–37.
7. M. Šyc, F. Georg Simon, J. Hykš, R. Braga, L. Biganzoli, G. Costa, V. Funari, and M. Grosso, *J. Haz. Mater.* 393 (2020) 122433.
8. P. Alvarenga, A.P. Gonçalves, R.M. Fernandes, A. de Varennes, E. Duarte, A.C. Cunha-Queda, and G. Vallini, *Waste Manag. Res.* 27 (2009) 101–111.
9. P. Alvarenga, C. Mourinha, M. Farto, T. Santos, P. Palma, J. Sengo, M.-C. Morais, and C. Cunha-Queda, *Waste Manag.* 40 (2015) 44–52.
10. R. Clemente, E. Arco-Lázaro, T. Pardo, I. Martín, A. Sánchez-Guerrero, F. Sevilla, and M.P. Bernal, *Chemosphere* 223 (2019) 223–231.
11. P. Alvarenga, A.P. Gonçalves, R.M. Fernandes, A. de Varennes, G. Vallini, E. Duarte, and A.C. Cunha-Queda, *Chemosphere* 74 (2009) 1292–1300.
12. A. Burges, I. Alkorta, L. Epelde, and C. Garbisu, *Int. J. Phytorem.* 20 (2018) 384–397.
13. C. Mastretta, S. Taghavi, D. Van Der Lelie, A. Mengoni, F. Galardi, C. Gonnelli, et al., *Int. J. Phytoremediat.* 11 (2009) 251–267.
14. Y. Ma, M. Prasad, M. Rajkumar, and H. Freitas, *Biotechnol. Adv.* 29 (2011) 248–258.
15. V. Göhre and U. Paszkowski, *Planta* 223 (2006) 1115–1122.

16. C.S. Seth, *Bot. Rev.* 78 (2012) 32–62.

17. H. Ali, M. Naseer, and M.A. Sajad, *Int. J. Environ. Sci.* 2 (2012) 1459–1469.

18. B.H. Robinson, E. Lombi, F.J. Zhao, and S.P. McGrath, *New Phytol.* 158 (2003) 279–285.

19. D.E. Salt, R.D. Smith, and I. Raskin, *Annu. Rev. Plant Phys.* 49 (1998) 643–668.

20. N. Rascio and F. Navari-Izzo, *Plant Sci.* 180 (2011) 169–181.

21. A. van der Ent, A.J. Baker, R.D. Reeves, A.J. Pollard, and H. Schat, *Plant Soil* 362 (2013) 319–334.

22. S.P. McGrath and F.-J. Zhao, *Curr. Opin. Biotechnol.* 14 (2003) 277–282.

23. A.P. Marques, A.O. Rangel, and P.M. Castro, *Crit. Rev. Env. Sci. Technol.* 39 (2009) 622–654.

24. A. Baker and R. Brooks, *Biorecovery* 1 (1989) 81–126.

25. S.T. Dushenkov, *Plant Soil* 249 (2003) 167–175.

26. J. Suman, O. Uhlik, J. Viktorova, and T. Macek, *Plant Sci.* 9 (2018) 1476.

27. B. He, X. Yang, W. Ni, Y. Wei, X. Long, and Z.J. Ye, *Integr. Plant Biol.* 44 (2002) 1365–1370.

28. X.E. Yang, X.X. Long, W.Z. Ni, and C.X. Fu, *Chin. Sci. Bull.* 47 (2002) 1634–1637.

29. X.E. Yang, X.X. Long, H.B. Ye, Z.L. He, D. Calvert, and P. Stoffella, *Plant Soil* 259 (2004) 181–189.

30. M. Arshad, M. Saleem, and S. Hussain, *Trends Biotechnol.* 25 (2007) 356–362.

31. B.R. Glick, *Biotechnol. Adv.* 28 (2010) 367–374.

32. V. Göhre and U. Paszkowski, *Planta* 223 (2006) 1115–1122.

33. T. Vamerali, M. Bandiera, and G. Mosca, *Environ. Chem. Lett.* 8 (2010) 1–17.

34. M. Wu, Q. Luo, S. Liu, Y. Zhao, Y. Long, and Y. Pan, *Ecotoxicol. Environ. Saf.* 162 (2018) 35–41.

35. P. Zhuang, Q.W. Yang, H.B. Wang, and W.S. Shu, *Water Air Soil Pollut.* 184 (2007) 235–242.

36. J.-L. Wei, H.-Y. Lai, and Z.-S. Chen, *Ecotoxicol. Environ. Saf.* 84 (2012) 173–178.

37. M.N.V. Prasad, *Heavy Metal Stress in Plants: From Biomolecules to Ecosystems*, Ed. M.N.V. Prasad, 2nd ed., pp. 345–391, Springer, 2004.

38. J.L. Schnoor, *Phytoremediation of Toxic Metals: UsingPlants to Clean Up the Environment*, Ed. I. Raskin and B.D. Ensley, pp. 133–150, John Wiley and Sons, Inc., 2000.

39. A. Williamson and M.S. Johnson, Reclamation of Metalliferous Mine Wastes. In *Effect of Heavy Metal Pollution on Plants. Pollution Monitoring Series*, Ed. N.W. Lepp, Vol. 2, Springer, 1981, https://doi.org/10.1007/978-94-009-8099-0_6

40. J. Vymazal, *Ecol. Eng.* 35 (2009) 1–17.

41. W. Zorrig, M. Rabhi, S. Ferchichi, A. Smaoui, and C. Abdelly, *J. Arid Land Studies* 22 (2012) 299–302.

42. R.L. Chaney, J.S. Angle, C.L. Broadhurst, C.A. Peters, R.V. Tappero, and D.L. Sparks, *J. Environ. Qual.* 36 (2007) 1429–1443.

43. R.R. Brooks, M.F. Chambers, L.J. Nicks, and B.H. Robinson, *Trends Plant Sci.* 3 (1998) 359–362.

44. S. Balzano, A. Sardo, M. Blasio, T. Chahine, F. Dell'Anno, C. Sansone, and C. Brunet, *Front. Microbiol.* 11 (2020) 517.

45. M. Laghlimi, B. Baghdad, H. El Hadi, and A. Bouabdli, *Open J. Ecol.* 5 (2015) 375–388.

46. M. Sabir, E. Waraich, K. Hakeem, O. Munir, H. Ahmad, and M. Shahid, *Chapter: Phytoremediation: Mechanisms and Adaptations*, Ed. R. Khalid, M. Hakeem, M. Sabir, O. Munir, and M. Ahmet, Academic Press, Elsevier, 2014.

47. M. Macnair, V. Bert, S. Huitson, P. Saumitou-Laprade, and D. Petit, *Proc. Biol. Sci. Royal Soc.* 266 (1999) 2175–2179.

48. S. Thakur, L. Singh, Z.A. Wahid, M.F. Siddiqui, S.M. Atnaw, and M.F.M. Din, *Environ. Monit. Assess.* 188 (2016) 206.

49. H. Lodish, A. Berk, and S.L. Zipursky, et al., *Molecular Cell Biology*, 4th ed., W. H. Freeman, 2000, Section 21.2, The Action Potential and Conduction of Electric Impulses.
50. R. Hedrich, *Physiol. Rev.* 92 (2012) 1777–1811.
51. D. Nelson and M. Cox, *Lehninger Principles of Biochemistry*, p. 403, W.H. Freeman, 2013.
52. R. Hedrich, *Physiol. Rev.* 92 (2012) 1777–1811.
53. D.A. Doyle, J. Morais Cabral, R.A. Pfuetzner, A. Kuo, J.M. Gulbis, S.L. Cohen, B.T. Chait, and R. MacKinnon, *Science* 280 (1998) 69–77.
54. Z. Lu, A.M. Klem, and Y. Ramu, *Nature* 413 (2001) 809–813.
55. R. MacKinnon, *FEBS Lett.* 555 (2003) 62–65.
56. B. Alberts, A. Johnson, J. Lewis, et al., *Molecular Biology of the Cell*, 4th ed., Garland Science, 2002, Ion Channels and the Electrical Properties of Membranes.
57. Y. Zhou, J.H. Morais-Cabral, A. Kaufman, and R. MacKinnon, *Nature* 414(6859) (2001) 43–48.
58. N. Merkl, R. Schultze-Kraft, and C. Infante, *Environ. Poll.* 138 (2005) 86–91.
59. A.A. Dalvi and S. Bhalerao, *Ann. Plant Sci.* 2 (2013) 362–368.
60. W.A. Peer, I.R. Baxter, E.L. Richards, J.L. Freeman, and A.S. Murphy, Phytoremediation and Hyperaccumulator Plants, In *Molecular Biology of Metal Homeostasis and Detoxification*, pp. 299–340, Springer, 2005.
61. G. Dal Corso, E. Fasani, A. Manara, G. Visioli, and A. Furini, *Int. J. Mol. Sci.* 20 (2019) 3412.
62. M. Guerinot, *Biochimica et Biophysica Acta* 1465 (2000) 190–198.
63. T. Mizuno, K. Usui, K. Horie, S. Nosaka, N. Mizuno, and H. Obata, *Plant Physiol. Biochem.* 43 (2005) 793–801.
64. A.G.L. Assunção, P. Da Costa Martins, S. De Folter, R. Vooijs, H. Schat, and M.G.M. Aarts, *Plant Cell Environ.* 24 (2001) 217–226.
65. K.B. Axelsen and M.G. Palmgren, *Plant Physiol.* 126 (2001) 696–706.
66. L.E. Williams and R.F. Mills, *Tr. Plant Sci.* 10 (2005) 491–502.
67. M. Hanikenne and D. Baurain, *Fron. Plant Sci.* 4 (2014) 544.
68. F. Verret, A. Gravot, P. Auroy, N. Leonhardt, P. David, L. Nussaume, A. Vavasseur, and P. Richaud, *FEBS Lett.* 576 (2004) 306–312.
69. J.L. Gustin, M.J. Zanis, and D.E. Salt, *BMC Evol. Biol.* 11 (2011) 76.
70. A.-G. Desbrosses-Fonrouge, et al., *FEBS Lett.* 579(19) (2005) 4165–4174.
71. M.W. Persans, K. Nicman, and D.E. Salt, *Proc. Nat. Acad. Sci. U.S.A.* 98 (2001) 9995–10000.
72. J. Lee, R.D. Reeves, R.R. Brooks, and T. Jaffré, *Phytochem.* 16 (1977) 1503–1505.
73. U. Krämer, J. Cotter-Howells, J. Charnock, et al., *Nature* 379 (1996) 635–638.
74. J. Hall, *J. Exp. Bot.* 53 (2002) 1–11, https://doi.org/10.1093/jexbot/53.366.1.
75. A.R. Memon and P. Schröder, *Environ. Sci. Pollut. R* 16 (2009) 162–175.
76. W.H. Ernst, J. Verkleij, and H. Schat, *Acta Bot. Neerl.* 41 (1992) 229–248, https://doi.org/10.1111/j.1438-8677.1992.
77. A. Manara, *Plants and Heavy Metals*, Ed. A. Furini, pp. 27–53, Springer, 2012.
78. S.S. Sharma and K.-J. Dietz, *J. Exp. Bot.* 57 (2006) 711–726.
79. D.K. Gupta, H. Vandenhove, and M. Inouhe, *Heavy Metal Stress in Plants*, Ed. D.K. Gupta, F.J. Corpas, and J.M. Palma, pp. 73–94, Springer, 2013.
80. G. Sarret, P. Saumitou-Laprade, V. Bert, O. Proux, J.L. Hazemann, A. Traverse, M.A. Marcus, and A. Manceau, *Plant Physiol.* 130 (2002) 1815–1826.
81. J.R. Domínguez-Solís, M.C. López-Martín, F.J. Ager, M.D. Ynsa, L.C. Romero, and C. Gotor, *Plant Biotechnol. J.* 2 (2004) 469–476.
82. S.B. Roy and A.J. Bera, *Environ. Biol.* 23 (2002) 433–435.
83. V. Rai, *Biol. Plantarum* 45 (2002) 481–487, https://doi.org/10.1023/A:1022308229759.
84. Y.-P. Tong, R. Kneer, and Y.-G. Zhu, *Trends Plant Sci.* 9 (2004) 7–9.

85. V. Sheoran, A. Sheoran, and P. Poonia, *Crit. Rev. Env. Sci. Technol.* 41 (2011) 168–214.

86. S. Eapen and S.D'Souza, *Biotechnol. Adv.* 23 (2005) 97–114.

87. D.K. Gupta, F.T. Nicoloso, M.R.C. Schetinger, L.V. Rossato, L.B. Pereira, G.Y. Castro, et al., *J. Hazard. Mater.* 172 (2009) 479–484.

88. M. Jozefczak, T. Remans, J. Vangronsveld, and A. Cuypers, *Int. J. Mol. Sci.* 13 (2012) 3145–3175.

89. G. Dal Corso, E. Fasani, A. Manara, G. Visioli, and A. Furini, *Int. J. Mol. Sci.* 20 (2019) 3412.

90. A.S. Davis, P. Prakash, and K. Thamaraiselvi, Bioremediation and Sustainable Technologies for Cleaner Environment, In *Environmental Science*, Ed. M. Prashanthi, pp. 13–33, Springer International Publishing AG, 2017.

91. B. Sattelmacher, *New Phytol.* 149 (2001) 167–192, http://sci-hub.tw/10.1046/j.1469-8137.2001.00034.x.

92. A.G. Roberts and K.J. Oparka, *Plant Cell Environ.* 26 (2003) 103–124.

93. J.G. Burken and J.L. Schnoor, *J. Environ. Sci.* 122(11) (1996) 958–963.

94. D.J. Glass, Phytoremediation of Toxic Metals, In *Using Plants to Clean Up the Environment*, Ed. I. Raskin and B.D. Ensley, pp. 15–31, John Wiley and Sons, Inc., 2000.

95. S. Kalve, B.K. Sarangi, R.A. Pandey, and T. Chakrabarti, *Curr. Sci. India* 100 (2011) 888–894.

96. M. Srivastava, L.Q. Ma, and J.A.G. Santos, *Sci. Total Environ.* 364 (2006) 24–31.

97. M. Sakakibara, Y. Ohmori, N.T.H. Ha, S. Sano, and K. Sera, *Clean Soil Air Water* 39 (2011) 735–741.

98. K. Peng, C. Luo, W. You, C. Lian, X. Li, and Z. Shen, *J. Hazard Mater.* 154 (2008) 674–681.

99. L. Buendía-González, J. Orozco-Villafuerte, F. Cruz-Sosa, C. Barrera-Díaz, and E. Vernon-Carter, *Bioresour. Technol.* 101 (2010) 5862–5867.

100. R. Kucharski, A. Sas-Nowosielska, E. Małkowski, J. Japenga, J. Kuperberg, M. Pogrzeba, et al., *Plant Soil* 273 (2005) 291–305.

101. J. Wang, X. Feng, C.W. Anderson, Y. Xing, and L. Shang, *J. Hazard Mater.* 221 (2012) 1–18.

102. L. Rodriguez, F. Lopez-Bellido, A. Carnicer, and V. Alcalde-Morano, *Fresen. Environ. Bull.* 9 (2003) 328–332.

103. A. Sas-Nowosielska, R. Galimska-Stypa, R. Kucharski, U. Zielonka, E. Małkowski, and L. Gray, *Environ. Monit. Assess.* 137 (2008) 101–109.

104. R.L. Chaney, C.L. Broadhurst, and T. Centofanti, *Soils*, Ed. P.S. Hooda, pp. 311–352, John Wiley and Sons Inc., 2010.

105. G. Koptsik, *Eurasian Soil Sci.* 47 (2014) 923–939.

106. S. Ebbs, M. Lasat, D. Brady, J. Cornish, R. Gordon, and L. Kochian, *J. Environ. Qual.* 26 (1997) 1424–1430.

107. J. Vangronsveld, R. Herzig, N. Weyens, J. Boulet, K. Adriaensen, A. Ruttens, et al., *Environ. Sci. Pollut. R* 16 (2009) 765–794.

108. T. Vamerali, M. Bandiera, and G. Mosca, *Environ. Chem. Lett.* 8 (2010) 1–17.

109. A. Kayser, K. Wenger, A. Keller, W. Attinger, H. Felix, S. Gupta, et al., *Environ. Sci. Technol.* 34 (2000) 1778–1783.

110. P. Tlustoš, J. Száková, J. Hrubı, I. Hartman, J. Najmanová, J. Nedìlník, et al., *Plant Soil Environ.* 52 (2006) 413–423.

111. R. Herzig, E. Nehnevajova, C. Pfistner, J.-P. Schwitzguebel, A. Ricci, and C. Keller, *Int. J. Phytoremediat.* 16 (2014) 735–754.

112. N. Sarwar, S.S. Malhi, M.H. Zia, A. Naeem, S. Bibi, and G. Farid, *J. Sci. Food Agric.* 90 (2010) 925–937.

113. E.P. Brewer, J.A. Saunders, J.S. Angle, R.L. Chaney, and M.S. Mcintosh, *Theor. Appl. Genet.* 99 (1999) 761–771.

114. E. Nehnevajova, R. Herzig, G. Federer, K.-H. Erismann, and J.-P. Schwitzguébel, *Int. J. Phytoremediat.* 9 (2007) 149–165.
115. A. Berken, M.M. Mulholland, D.L. Leduc, and N. Terry, *Crit. Rev. Plant Sci.* 21 (2002) 567–582.
116. A. Koźmińska, A. Wiszniewska, E. Hanus-Fajerska, and E. Muszyńska, *Plant Biotechnol. Rep.* 12 (2018) 1–14.
117. D. Mani and C. Kumar, *Int. J. Environ. Sci. Technol.* 11 (2014) 843–872.
118. N. Das, S. Bhattacharya, and M.K. Maiti, *Plant Physiol. Biochem.* 105 (2016) 297–309.
119. G. Wu, H. Kang, X. Zhang, H. Shao, L. Chu, and C.J. Ruan, *Hazard Mater.* 174 (2010) 1–8.
120. E. Fasani, A. Manara, F. Martini, A. Furini, and G. DalCorso, *Plant Cell Environ.* 41 (2018) 1201–1232.
121. B.R. Glick, *Microbiol. Res.* 169 (2014) 30–39.
122. M. Arshad, M. Saleem, and S. Hussain, *Trends Biotechnol.* 25 (2007) 356–362.
123. V. Göhre and U. Paszkowski, *Planta* 223 (2006) 1115–1122.
124. A. Verma, A. Roy, N. Bharadvaja, Chapter 13 – Remediation of Heavy Metals Using Nanophytoremediation, In *Advanced Oxidation Processes for Effluent Treatment Plants*, Ed. M.P. Shah, pp. 273–296, Elsevier, 2021, ISBN 9780128210116, https://doi.org/10.1016/B978-0-12-821011-6.00013-X.
125. Y. Cao, S. Zhang, Q. Zhong, G. Wang, X. Xu, T. Li, L. Wang, Y. Jia, and Y. Li, *Ecotoxicol. Environ. Saf.* 162 (2018) 464–473.
126. A.M. Ahmed, M. Ali, and A.H. Noor, *Am. J. Mater. Sci.* 6 (2016) 105–114.
127. M. Alimohammady, M. Jahangiri, F. Kiani, and H. Tahermansouri, *Res. Chem. Intermed.* 44 (2018) 69–92.
128. S. Luo, T. Lu, L. Peng, J. Shao, Q. Zeng, and J. Gu, *J. Mater. Chem.* 2 (2014) 15463–15472.
129. N. Mueller, J. Braun, J. Bruns, M. Cernik, P. Rissing, D. Rickerby, and B. Nowack, *Environ. Sci. Pollut. Res.* 19 (2012) 550–558.
130. D. Huang, X. Qin, Z. Peng, Y. Liu, X. Gong, G. Zeng, C. Huang, M. Cheng, W. Xue, X. Wang, et al., *Ecotoxicol. Environ. Saf.* 153 (2018) 229–237.
131. K. Chekroun and M. Baghour, *J. Mater. Environ. Sci.* 4 (2013) 873–880.
132. C.C. Carias, J.M. Novais, and S. Martins-Dias, *Water Sci. Technol.* 56(3) (2007) 263–269.
133. D.A. Yaseen and M. Scholz, *Environ. Sci. Pollut. Res.* 25 (2018) 1980–1997.
134. B. Can-Terzi, A.Y. Goren, H.E. Okten, and S.C. Sofuoglu, *Environ. Technol. Innov.* 22 (2021) 101432.
135. D. Nisha and T. Emilia, *Biochem. Eng. J.* 115 (2016) 23–29.
136. E.A.S. Almaamary, S. Abdullah, H. Abu Hasan, R. Ab Rahim, and M. Idris, *Mal. J. Anal. Sci.* 21 (2017) 182–187.
137. K.A. Tan, N. Morad, and J. Ooi, *Int. J. Environ. Sci. Develop.* 7 (2016) 724–728.
138. A. Kagalkar, M. Jadhav, V. Bapat, and S. Govindwar, *Biores. Technol.* 102 (2011) 10312–10318.
139. X. Zhou and X. Xiang, *Procedia Environ. Sci.* 18 (2013) 540–546.

6 Chemical precipitation, coagulation, and flocculation

Triad of great versatility

6.1 INTRODUCTION

In meteorology, the term precipitation refers to water that falls back on the earth's surface due to condensation. But in science it is the formation of particles or powders from solutions under different conditions. This phenomenon is not new to man. Some such problems encountered in daily life due to precipitation are clogging of water pipes at home [1] or even the formation of kidney stones [2]. Thus, suspended colloidal particles are bound to be more dangerous to human health [3,4]. However, this concept can be used for the removal of toxic species from solution.

Chemical precipitation is a separation technique based on the solubility of the components present in a mixture. According to the IUPAC, solubility can be defined as the analytical composition of a saturated solution expressed as the proportion of a designated solute in a designated solvent [5]. The solubility of a solute lies in the range of being insoluble (poorly soluble like silver chloride in water) to infinitely soluble without limit (miscible like ethanol in water) [6]. Solubility can be expressed in any one of the units such as molarity, molality, mole fraction, mole ratio, or mass (solute) per volume (solvent). According to *US Pharmacopeia* [7], the unit of solubility is defined as the mass parts of a solvent needed for dissolution of 1 mass part of a solute. The threshold limit of insolubility depends on the application of precipitation reaction [8] as solubility depends on various factors like ionic strength and pH of a solution and the reaction temperature. Therefore, solubility is defined as the maximum amount of a substance that dissolves in a given amount of solvent at a specified temperature. Hence, the fine-tuning of experimental conditions to render a solute insoluble is known as precipitation, and the insoluble solid formed is known as precipitate. The precipitate primarily exists as a well-dispersed suspension, which finally agglomerates and can be removed by sedimentation, centrifugation, or filtration. Different parameters like pH and temperature have a strong effect on precipitation. Chemical precipitation is the process by which a substance soluble in a solution can be converted into an insoluble form (precipitate) from a supersaturated solution. The clear solution present along with the precipitate is referred to as a supernatant or supernate. A supersaturated solution contains a particular species in concentrations greater than its solubility limit. Precipitation is a complex phenomenon resulting from the formation of conditions of supersaturation [9]. Three steps are involved in the process: nucleation (germination), crystal growth, and flocculation [10].

DOI: 10.1201/9781003442516-6

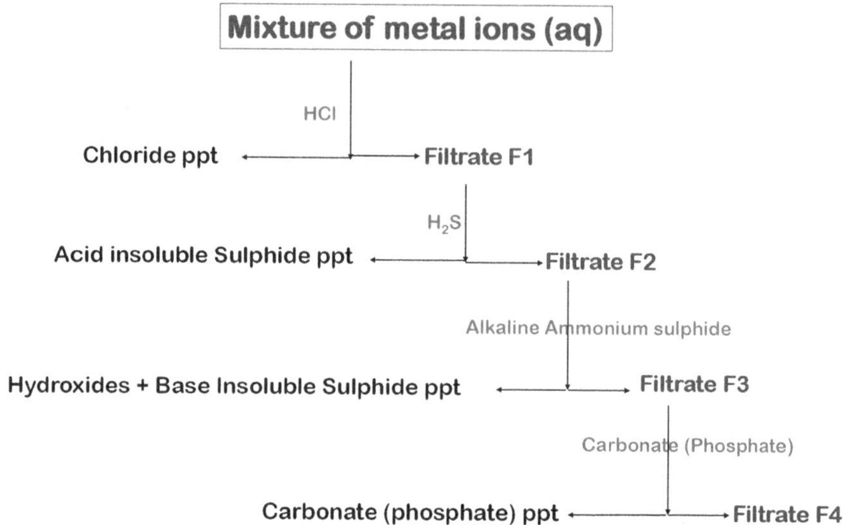

FIGURE 6.1   Schematic representation of separation of metal ions by precipitation

Nucleation is the stage at which the first germ appears due to condensation. The germ then aggregates until supersaturation of the solution occurs, and the solid is finally formed. Metal ions in the solution are precipitated as oxides, hydroxides, sulphides, carbonates, phosphates, etc. The precipitate is separated from the supernatant solution using methods like filtration, centrifugation, etc. Figure 6.1 presents a schematic representation of the precipitation of successive groups of metal ions with nearly similar chemical properties. The separation involves the chronological precipitation of the filtrate obtained in the previous step. In the first step, the aqueous solution containing a mixture of ions is treated with hydrochloric acid. The ions that form insoluble chloride salts are precipitated, while others are present in the filtrate. The filtrate is subsequently treated to obtain sequential separation using an appropriate precipitating agent. It is known that most of the metal chlorides are water-soluble, and the only ions that form precipitates are Ag^+, Pb^{2+}, and Hg_2^{2+}, which could be removed by chloride precipitation. The use of sulphide reagent results in the precipitation of metal sulphides of As^{3+}, Bi^{3+}, Cd^{2+}, Cu^{2+}, Hg^{2+}, Sb^{3+}, and Sn^{2+}, and hence their separation in the second stage. The addition of sodium hydroxide results in the hydroxo precipitation of Al^{3+} and Cr^{3+}, while the addition of alkaline sulphide precipitates most of the transition metal ions such as CO^{2+}, Fe^{2+}, Mn^{2+}, Ni^{2+}, and Zn^{2+}. The addition of carbonate or phosphate promotes the precipitation of alkaline earth metal ions such as Ca^{2+}, Mg^{2+}, Sr^{2+}, and Ba^{2+}. The filtrate obtained finally contains the alkali metal ions $\left(Li^+, Na^+, K^+, Rb^+, \text{and } Cs^+\right)$ and ammonium $\left(NH_4^+\right)$. Thus, precipitation is useful for the simple qualitative separation prior to analysis.

    Chemical precipitation by hydroxide addition is a popular protocol due to its simplicity, its low cost, and ease of controlling of experimental conditions. In the pH range of 8–11, the solubility of various metal hydroxides are minimum and so they can be removed easily. Lime is the most preferred choice of base [11]. There are different classes of reagents known as true metal precipitants, and the most common metal precipitant is dithiocarbamate (DTC; as sodium dimethyl or diethyldithiocarbamate, DDTC), which forms stoichiometric compounds which are formed as co-precipitates with hydroxides [12]. However, care should be taken to avoid large concentrations of DTC due to its degeneration into sulphates and nitrates, which can change the water quality. To avoid possible contamination, non-sulphur-containing precipitants like clay-based products, aluminium, and cationic polymers impregnated with iron and polymerized aluminium derivatives are gaining popularity. Clay-based products are basically coagulants that are generally used to treat organic pollutants.

Coagulation and flocculation are processes that have been used from ancient times to treat dirty water [13, 14]. These two processes can be utilized for the separation of finely dispersed colloidal particles having different characteristics such as charge, size, shape, and density. For obtaining stable dispersion, aggregation should be minimum. But upon the addition of a coagulant or flocculant, the particles aggregate due to destabilization of colloids. The mostly used reagents are alum, ferric sulphate, ferric chloride, etc. Coagulation by charge neutralization is accomplished either by reducing the zeta potential of colloids or by flooding the medium with an excess of oppositely charged ions [15]. Agglomeration is a sum of sequential steps of coagulation, flocculation, and sedimentation. It is very essential that each step must be completed for the successive stages to be completed and successful. Coagulation by charge neutralization results in the formation of microflocs, which upon gentle mixing become larger in size and are known as pinflocs. The continued increase of the floc size results in the formation of macroflocs that settle down and are separated as sedimentation. As the size of the floc increases, the mixing velocity and energy are decreased to prevent the tearing of the floc. The coagulant is chosen based on the nature of the chemical species to be removed. Inorganic coagulants such as aluminium and iron salts are used to neutralize suspended particles. Inorganic hydroxides are formed as small polymers, which enhance the microfloc formation. Inorganic coagulants are very economical and can remove microorganisms. The addition of coagulants results in a change in the pH of water. Long-chained polymer coagulants work over a wide pH range compared to inorganic coagulants. This is because a very small amount of polymer is used, and the water parameters are not affected. The suspended solid in fluid media is a heterogeneous hydrodynamic system that can be controlled by gravity, centrifugal force, buoyancy, pressure, and electric current. Sedimentation is controlled by gravity, while centrifugal force operates in a centrifugation system. When the particle is suspended in solution, the fluid (liquid) flows over the particle and a resistance is created. To reduce this, the flow of fluid is made uniform and smooth by using energy. Thus, the resistance developed and the energy required are dependent on the geometry of the solid particle. When flow velocity is low, laminar flow conditions prevail, leading to the formation of a well-defined layer around the particle to get a uniform flow. Increase in the fluid velocity makes the flow turbulent, and the boundary reduces. Thus by using various mathematical expressions, the settling velocities can be calculated. The resistance of surrounding water will also increase as the particle descends and finally becomes equal to accelerating forces. From this point, the particle starts falling down with a constant maximum velocity, known as terminal velocity. Sedimentation is the removal of suspended particles due to gravitational settling, and the clear liquid is decanted. The most common coagulants are the inorganic salts of Al and Fe. Both of them form insoluble gelatinous hydroxides that can be used to remove suspended particles. The following equations show the various chemical reactions that occur upon the addition of salts. It has been found that aluminium salt is suited for a pH range of 5–8, while iron salts work better in higher pHs of 8–10.

$$Al_2\left(SO_4\right)_3 + 3Ca\left(HCO_3\right)_2 \Leftrightarrow 2Al\left(OH\right)_3 + 3CaSO_4 + 6CO_2$$
$$Al_2\left(SO_4\right)_3 + 3Na_2\left(CO_3\right) + 3H_2O \Leftrightarrow 2Al\left(OH\right)_3 + 3Na_2SO_4 + 3CO_2$$
$$Al_2\left(SO_4\right)_3 + 6NaOH \Leftrightarrow 2Al\left(OH\right)_3 + 3Na_2SO_4$$

Different species can behave as coagulants, but each of them has its own merits and demerits [16]. Aluminium sulphate [alum: $Al_2(SO_4)_3 \cdot 18H_2O$] is the commonly used coagulant as it produces less sludge, is effective in mild pH conditions $(6.5-7.5)$, and is very easy to handle. But the main disadvantages are its applicability in a limited pH range and solubility in alkaline conditions. Sodium aluminate $(Na_2Al_2O_4)$ is very effective in hard waters and can be used in very small amounts. The main limitations are its cost and ineffectiveness in soft water. Polyaluminium chloride [PAC; $Al_{13}(OH)_{20}(SO)_4Cl_{15}$] results in a more denser floc compared to alum, and this settles faster. Despite these advantages, there are not much reports on the use of PAC. Ferric sulphate [$Fe_2(SO_4)_3$] is effective in both mildly acidic (pH 4–6) and mildly alkaline conditions $(8.8-9.2)$, but the water becomes alkaline due to solubilization. Ferric chloride $(FeCl_3.6H_2O)$ is effective for a wide pH range of 4–11, but dissolution hampers its use. Similarly, ferrous sulphate $(FeSO_4.7H_2O)$ is not quite sensitive to pH change, but its solubility leads to a change in pH of the water. Lime $(Ca(OH)_2)$ is very commonly used because it is effective and does not have solubilization issues, but it produces large amounts of sludge. Inorganic coagulants with many advantages are still less attractive as sludge production is high, and the sludge does not dry up rapidly. Therefore, pre-polymerized aluminium solutions (with variable degrees of polymerization) like PAC, polyaluminium sulphates (PAS), or polyaluminium chloro-sulphates (PACS) are used. The PAC have found applications due to their inherent advantages of being applicable over a wide *pH* range, being resistant to temperature and colloidal concentrations, and requirement of low coagulant dosage. These polymeric structures are produced from soluble anionic aluminium hydroxo complexes [$Al(OH)_{4-}$], formed due to hydrolysis at pH > 3. These polymerized coagulants are not as efficient as organic polyelectrolytes [16], whose higher molecular weight (MW) leads to better flocculation. However, their very high molecular weight might lead to viscous solution, and appropriate dilution is required [17]. The two important characteristics of polymers, namely, high charge density (for efficient charge neutralization) and low molecular weight (for easy diffusion around particles), make polymers suited as coagulants or flocculants [18, 19]. Organic polymers can be classified into low, medium, and high when their molecular weights are less than 10^5, 10^5–10^6, and greater than 10^6, respectively. The initial mixing is very essential for complete dispersion of the polymer. Flocculation is the second stage, which commences once coagulation is nearly complete. Sometimes bridging of two turbid particles to different chains of the polymer also occurs. For aggregation to occur, the mixing energy should be sufficient to ensure particle collision but not very high, resulting in the collapse of aggregates. Inorganic or organic flocculant aids are used to overcome the issues of aggregate disintegration. Polymerized silica is generally used to improve flocculation via composite polymerization (introduction of polymerized silica in the pre-polymerized metal solution) or copolymerization (introducing polymerized silica in the metal solution) [20]. The increasing concentration of monosilicic acid to levels greater than the solubility limit results in the precipitation of the polymer. It is seen that different silica species can form depending on various experimental parameters such as temperature, silica concentration, pH, other ions, and time of ageing [21]. The most important parameter that affects silica polymerization is the pH value as hydroxyl ions probably catalyse the reaction. However beyond pH of 9, the polymerization decreases, and with ageing, gel formation is observed [22]. The gel formation can be circumvented by amending the pH. Polyaluminium silicate chloride is superior to conventional or simple pre-polymerized coagulants [23]. Acrylamide-based polymers and copolymers like hydrolysed polyacrylamides (anionic, molecular weight: 104–107 g/mol) and polydiallyldimethyl ammonium chloride (poly-DADMAC; cationic, molecular weight: 104–107 g/mol) are useful. In general, organic polymers prove to be more useful as flocculant aids rather than as primary coagulants. Natural polymers (biopolymers) like pre-gelatinized starch derivatives, amine-treated cationic starches, guar gums, tannins, chitosan, and alginates behave as flocculants (bioflocculants) [25, 26].

It is seen that co-precipitation with inorganic precipitants is carried out along with colloidal precipitates (sulphides and hydroxides) having a large surface area [27]. Thus, co-precipitation has been studied with hydroxides of iron(III) and manganese(IV) to a greater extent and with that of Al, Be, La, and Zr to a lesser extent. Co-precipitation is used to separate and concentrate trace elements from very dilute solutions, such as natural water. Since the solubilities of metal hydroxides or sulphides are mainly governed by the pH value of the solution, it is known that many of the metal ions can be co-precipitated with iron hydroxide in the pH range of 9–10. However, different parameters like pH, quantity of precipitant, coexisting salts, precipitate ageing time, etc., have a profound effect on the yield. Co-precipitation can occur due to different mechanisms like an ion exchange reaction (between the hydrogen on the hydrated oxide surface and trace metal ions in the solution ($nSH— + Mm+ \rightleftharpoons knSnM—(m − n) + +nH + ¬$), sorption of metal ions and then hydroxide ions onto the surface of the metal oxide, or sorption of the hydroxo metal complex onto the surface. Dithiocarbamate (DTC) is very commonly used to achieve separation between heavy metal ions by precipitation. Clay-based products are primarily used as coagulant aids due to the excellent swelling ability of the material and the ease of functionalization.

Coagulation is carried out in a conventional clarification system [28], which contains a large tank with separate chambers for mixing and settling. The periods for coagulation (retention), flocculation, and settling are in the range of 3–5 minutes, 15–30 minutes, and 4–6 hours, respectively [29]. The first two steps are very crucial to achieve excellent clarification. Hence, the initial mixing is quite turbulent, ensuring complete contact between the coagulant and suspended particles. The flocculated particles settle down at the bottom, while clean water passes out to the circumferences onto clean weir. The accumulated sludge is scraped off. This horizontal basin clarifier can produce high-quality effluent in an upflow clarifier (when the water flow is in the upward direction). Air flotation systems can remove solids, oil, grease, and fibrous materials. Air flotation is the production of microscopic air bubbles, which enhance the natural tendency of some materials to float by recycling back a small part of the clarified flow. Dissolved air flotation (DAF) unit is a water treatment process that clarifies waste waters by the removal of suspended matter by dissolving air in the water or waste water under pressure. The pressurized air released at atmospheric pressure in a flotation tank basin forms tiny bubbles. These bubbles adhere to suspended matter, making it float on the surface, which is removed by skimming [30, 31]. In flotation, it is not essential to obtain a large quantity of heavy flocs to achieve efficient removal of solids. Therefore, it becomes easy to use DAF as both flocculation time and reagent dosages are low. The forces applied to sludge by bubble agglomerates enhance the efficiency of chemical use and reduce the volume of residuals to be treated.

Filtration is a separation process wherein suspension is passed through a porous membrane or medium [32]. The medium retains the solid particles on its surface or within its pores. The solid is known as filter, while filtrate is the fluid passing through the membrane. A filter is characterized by four features, namely, porosity, permeability, tortuosity, and connectivity. Porosity is the fraction of the medium with voids. The pores can be ordered or random. Conductivity is dependent on the porous structure. There are two modes of the passing of liquid through the filter: (i) gravity filter (liquid flows through the filter under the gravity) and (ii) vacuum filter (gravity is not sufficient to induce flow, and a vacuum is applied on the discharge side). Filtration along with clarification can remove suspended particles. The interaction between the particles and the filter medium determines the filtration mechanism. In practice, cake filtration is quite common. Cake formation refers to the deposition of a thin layer of residual particles on the surface of the filter medium [33]. The edifice of the cake formed (determined by cake porosity, mean particle size, size distribution, and particle-specific surface area and sphericity) and its resistance to liquid flow depend on the properties of suspension and filtration experimental conditions (coagulation rate, applied pressure, and temperature). The resistance to liquid flows leads to reduced filtration efficiency and increased operation time. In general, most of the cakes formed are compressible, and the rate of compressibility is inversely proportional to particle size. The presence of differently sized particles affects the cake structure from the time of formation until the end. This is because small particles may be transported from the top surface of the cake to near the filter or sometimes within the pores of the filter, resulting in pore clogging, and exhibit of an increased resistance to the flow. Dewatering is the process of washing the filter cake to effect clean separation of the solid and mother liquor [34]. In this, a clean fluid is passed through the cake to recover the residual liquids in the pores. However, this causes the structure of the cake to disintegrate. There are different types of setup of filtration medium. The assembly of plates and frames is the most common arrangement. In this setup, there is an opening at one corner for the introduction of feed slurry, and the hollow centre has an auxiliary channel. As the feed flows into the setup, the formation of cake in the hollow centre commences. The filtrate streams all the way through the medium onto the grooves present on the filter plate and finally through an outlet channel in each configuration. This continues until the flow of the filtrate decreases below a practical level (or pressure builds up too much). Upon termination of filtration, the wash liquid is passed in the same mode to wash the cake. However, for better performance, special wash plates arranged within the configuration are used. The setup is such that every other plate is a wash plate. During washing, the outlets of wash plates are closed, and the washed liquid is sent through a special inlet channel. The cake is removed manually once the entire process is over. An alternate but familiar setup is the one which uses a recessed-plate press, in which the cake gets collected within the recesses in the plates and no frames are used. The feed is sent in at the centre and removed from the corners of the configuration. The substitution of the slurry by the wash liquid is the protocol adopted to wash the cake. The vertical plate filter is a simple but widely used design, which is compact, economical, and workable, and thus finds extensive use in the industry. The design needs labour and is costly due to replacement of the filter cloth. The filters are classified based on the method of capture, i.e. exclusion at the filter surface (straining and deposition) within the media (in-depth filtration). Strainer is a simple thin physical barrier (metal/plastic) used at the inlet of the water treatment process to preclude large contaminants (leaves, fish, and coarse detritus) with an interspacing of 1–10 cm. Microstrainers with lower interspacing are used to remove fine silt and algae. Metallic microstrainers can be used as a perforated sheet or woven wire. However, a plastic one is in the form of a woven or fused grid. The straining medium for wastewater treatment consists of a bundle of twisted fibres. A filter used normally for point-of-use treatment is disposed of after use. Cartridge filter consists of a disposable cartridge, which contains a porous and non-compressible material wound on a cylindrical support. It finds utility in small-scale applications like domestic point-of-use water treatment. In-depth granular media filtration with filters containing sand or crushed anthracite is used in municipal water treatment. Slow-sand filtration and rapid gravity or pressure filtration are two modes of operation using granular media filters. Pre-coat filtration uses a thin inert medium coated with loose fibres or powders [35]. Slow-sand filters reduce the filtration rate to avoid the capture of contaminants deep within the bed. It involves straining and capture within top 20 cm of the sand. The accumulated fragments, known as schmutzdecke [36], contribute to biological treatment of water. However, it is essential that the water entering the filters must not contain any disinfectant that might interrupt the biological activity of schmutzdecke. It is also essential to remove the floc particles formed to reduce the build-up of resistance to the flow. Rapid gravity and pressure filters are used under gravity (rapid gravity filtration) or in pressure (pressure filtration)-driven processes [37]. The basic mechanisms of particle removal are fundamentally the same in both gravity and pressure modes and differ only in the flow. Different granular media filters with unique applications have their own merits and demerits. Membrane filters using cloth or fabric in ancient times and synthetic membranes in modern times have become very popular.

Electrocoagulation is the means of using electric current to subvert colloidal particles in waste water [38] by simultaneous oxidation and reduction processes.

Universally, the anodes are fabricated from iron. It has been observed that many hydrolysed species are formed depending on the metal ion concentration and pH of the solution.

$$\text{Anodic Reaction}: Fe_s \rightarrow Fe^{2+}_{aq} + 2e^-$$

$$\text{Cathodic Reaction}: 2H_2O + 2e^- \rightarrow H_{2(gas)} + 2OH^-$$

$$\text{Overall Reaction}: Fe_s + 2H_2O \rightarrow Fe^{2+}_{aq} + H_{2(gas)} + 2OH^-_{aq}$$

Electroflotation (EF) is a process wherein bubbles of hydrogen and oxygen are produced by electrolysis at the cathode and anode, respectively. These bubbles then collide with the suspended particles in the waste water and make them float. The water treatment is achieved by skimming floating particles containing the suspended solid.

Crystallization is a separation process based on the limited solubility of a compound in a solvent at a certain temperature, pressure, etc. [39]. The change in conditions leading to decreased solubility leads to the formation of a crystalline solid. Crystallization has been used for ages for the production of salt, but the mechanisms of nucleation and crystallization of complex solutes are still intangible and also there is a lack of proper facility for carrying out the same. However, the increased requirement for excellent product quality deems crystallization as an attractive technique. The theoretical background of crystallization being a process of formation of crystalline solids with highly regular internal structures (crystal lattice) from liquid, gas, or amorphous solids is well known. The formation of such a highly ordered structure excludes impurities from the lattice, leading to high purity. The important development during crystallization is lucid from the rates of nucleation B ($\#/m^3s$) and growth G (m/s). The nucleation rate B is the number of new crystals formed per unit time and volume of suspension, while the rate of growth G is the rate at which the size of the crystals increases. Supersaturation influences the rate of the two reactions. Once crystal formation has commenced, secondary nucleation occurs at lower levels of supersaturation and is dependent on the stirring rate. Stable nuclei are formed only if the increase in interfacial energy (from solid–liquid interface formation) is counterbalanced by the energy decrease due to the release during the formation of a thermodynamically stable solid. The unpredictable nature of nucleation, coalesced with its effect on the crystal size distribution, makes the process uncontrolled, and this limitation is overcome by seeded crystallization (addition of small, previously produced crystals) for industrial applications. In cooling crystallization, the decrease in temperature can lead to supersaturation, and cooling has a great effect on the end product. When the solution and surroundings have a huge temperature difference, the cooling rate will be high, with a very sharp change in the initial stages which becomes least at the end of crystallization. Thus, the seed crystals cannot grow fast enough to consume the supersaturation, resulting in products with a wide distribution of crystal sizes. If the cooling is carried out in a linear manner, the initial supersaturation is reduced, and this can result in nucleation. Hence, it is of immense importance to maintain the conditions of supersaturation [10, 40, 41].

6.2 APPLICATIONS FOR THE REMOVAL OF METAL IONS AND DYES

The efficiency chemical precipitation depends on different experimental factors such as pH, precipitate, mixing time and speed, sludge volume, and complexing agents. The complete removal of Cr from tannery effluents using alkaline precipitation is difficult due to the presence of organic ligands in waste solutions. The use of hydroxides of calcium $[Ca(OH)_2]$ and sodium $(NaOH)$, and calcium magnesium carbonate $[CaMg(CO_3)_2]$ is found to be useful for the removal of chromium from tannery waste water [42, 43]. It has been observed that more than 99% of Cr could be removed with minimum sludge production upon using a combination of NaOH and $Ca(OH)_2$ [44]. Another study reported that MgO is a better precipitating agent than NaOH or $Ca(OH)_2$ at pH 9.8–10.3 [45], but a combination of MgO and $Ca(OH)2$ is very effective [46]. Heavy metal precipitation by hydrolysis is found to be quite effective using common precipitants [47]. The formation of bridged compounds was attributed to improved efficiency of a mixture of CaO and MgO compared to NaOH [48, 49]. Under acidic conditions, secondary precipitates contributing to sludge formation are formed upon the use of calcium hydroxide precipitants. The formation of secondary waste makes it necessary to develop a method to dispose the waste, thereby making the whole process economically unfavourable. In sulphide precipitation [50–52], neutralization of acidic effluents is essential to avoid the generation of toxic H_2S gas [50]. Metal sulphides have lower solubility than hydroxides, resulting in complete precipitation, and show improved stability over a wide pH range. Sulphide precipitation has more advantages compared to hydroxides due to lower residual metal concentration, less interference from chelating agents (present in waste water), better selectivity, and faster reaction rate. Moreover, sulphide sludges dry better and can be treated to recover metal ions. However, the main disadvantages are the high cost of the reaction and the possible formation of toxic H_2S gas (acidic conditions or when the molar concentrations of sulphide ions are in excess of metal ions). The metal ions can be precipitated as metal carbonates using reagents like Na_2CO_3 and $CaCO_3$ [53]. The majority of carbonates have a greater solubility than metal hydroxides. Separation of metal ions as their phosphates [54, 55] has also been reported but not used extensively. The majority of phosphates have a lower solubility than metal hydroxides. Various other precipitants have also been synthesized and used for separation. The most commonly used reagent is DTC [56, 57]. Precipitants can be classified as macromolecular trapping agents and small molecular precipitants [58]. Xanthate, polymerized ethylene dichloride–ammonia dithiocarbamates, or polydialkyl amine dithiocarbamates are macromolecular trapping agents, while DDTC and dimethyldithiocarbamate (DMTC) are small molecular precipitants [59, 60]. Reagents like 2,4,6-trimercapto-1,3,5-triazine (TMT) trisodium salt nonahydrate and Thio-Red (potassium/sodium thiocarbonate, STC) are found to be useful for the precipitation of different toxic species [61]. Co-ordination polymerization precipitation involves the use of a co-ordinating ligand to form the precipitate [62]. Novel ligands based on pyridine thiol have been found to be suited for the removal of toxic species [63]. Fenton reaction is usually used to improve the removal efficiency of chemical precipitation methods [64–68]. The removal of dyes is carried out using precipitation. Alkaline white mud (AWM) is found to be useful for the removal of acid blue 80 [69]. It has been found that precipitation is the prominent mechanism for the removal of high concentrations of dye. The use of chlorides of divalent magnesium and manganese has been reported for the removal of synthetic textile waste waters containing the azo-dye pigment Levafix Brilliant Blue EBRA [70]. It has been observed that chloride of manganese was more superior to that of magnesium as a coagulant. The coagulation occurs due to the formation of hydroxide species. It was established that brucite[$Mg(OH)_2$] particles were formed when applying $\left(MgCl_2\right)$, whereas a mixture of feitknechite $\left(\beta - MnOOH\right)$ and hausmannite $\left(Mn_3O_4\right)$ is obtained when using $MnCl_2$. The azo-dye pigment sorbs onto the surface of precipitating phases, and charge neutralization occurs leading to the formation of aggregates. Leaching solutions of white mud could remove different dyes like acid orange II, reactive light yellow K-6G, reactive bright red K-2BP, and direct yellow R within 90 seconds [71]. Water quality was maintained by using precipitation. Separation of different metal ions with precipitation has been reported [72–75]. Table 6.1 presents a list of the radionuclides separated using different reagents. Separation is achieved by altering the acidity or alkalinity of a solution.

Coagulation and flocculation find applications in the treatment of dye-loaded waste water. Calcium carbonate and hydroxide green synthesized from gastropod shells and in situ hybridization of dyes (methylene blue and Congo red) have been impregnated into the growing calcium salt derivatives [76]. It has been observed that precipitation of methylene blue was about 80%, while that of Congo red was 98%. The process was independent of pH, initial dye concentration, presence of anions, and ionic strength. The presence of anions increased the quantity of sludge produced. To

---

TABLE 6.1

List of precipitants for radionuclide separation

Precipitants		Trace Elements
Hydroxide	Al	Cr(III), Mo, W, Ce, Eu, La, Lu, Nd, Sm, Tb, Th, Tm, Yb, Th, U
	Be	As, P
	Bi(III) + In(III)	Co, Cu, Fe, Mg, Ni
	Fe(III)	Cd, Co, Eu, Mo, Cr(III), (VI), As, Ge, P, Sb, Se, Te, W, Cu, Mn, Ni, Pb, Zn, V, Ag, Ce, Cr, Cs, Er, Gd, La, Mn, Rb, Sr, Yb, Zn
	Ga(III)	Al, As, Cd, Co, Cr, Cu, Fe, La, Mn, Ni, Pb, Ti, Zn
		Cd, Co, Cr, Cu, Fe, Mn, Ni, Pb
	La(III)	As, Bi, Sb, Se, Te
		Co, Fe, Mn, Ni, Zn
	Mn(IV)	Mo Ga, As, Se
Sulphides	Pb(II)	Cu
	Cd(II)	Co, Cr, Cu, Mn, Zn
Sulphate	Na	Cr(III)/Cr(VI)
Fluoride	Ca(II)	U, Th

overcome this, emulsion splitting electrolysis (similar to electroflotation) in combination with classical liquid–liquid separation operation is reported [77]. The removal of cibacron red FN-R using an inorganic–organic composite polymer through a flocculation system was found to be dependent on polymer dosage, solution pH, dye concentration, agitation speed, time, and nature of coagulants [78]. The use of ferric chloride as coagulants for the removal of disperse red dye [79–81] was found to be more efficient than the use of aluminium sulphate, with an efficiency of greater than 90%. The presence of high salt content decreased the efficiency. Natural occurring biodegradable coagulants like chitosan, surjana seed powder, and maize seed powder were found to be very effective [82]. Bittern, a by-product of solar salt production, contains high concentrations of magnesium ions, enabling its use as coagulants in the removal of indigo blue dye [83]. Acid cyanine 5R and direct violet N dyes were removed by flocculation using epichlorohydrin–dimethylamine [84].

6.3 CONCLUSIONS

Chemical precipitation is used for the uptake of pollutants and separation of the products formed. This process is quite simple, efficient, and also cost effective. It can be applied to very high concentration levels of toxic species, but it is not highly selective. However, this process also has some disadvantages. The first disadvantage is the large quantity of chemicals consumed in the process. It is not useful for low concentration levels of toxic species. It is also not suitable if the metal ions are not free in the solution. This process leads to sludge formation, which needs extra efficiency to be handled and disposed. The additional steps make the cost higher.

Coagulation and flocculation are processes used for the removal of pollutants. This combination process is simple and cost-effective. A wide range of chemicals can be used as coagulants, and the sludge formed can be easily removed from the water body. This process can be used to reduce the chemical and biochemical oxygen demands and also the bacterial activity in water. However, this process also has certain limitations. It needs large amounts of chemicals, and the effluent parameters can be altered. There is an increase in the sludge volume, which needs to be managed and treated, resulting in an increase in the cost of the process. It is not suited for many species, especially arsenic.

Flotation or froth flotation is an integrated physicochemical separation process. Different types of collectors (ionic and non-ionic) are used for the efficient removal of small particles which would normally not settle down quite fast. It is used for the primary clarification method and is found to be selective. However, the process requires very high initial capital cost and also has high energy requirements. The maintenance and operation costs are also significant, and the choice of chemicals is critical to obtain proper forth characteristics, and the process is sensitive to pH.

REFERENCES

1. C.G.E.M. van Beek, C.H.M. Hofman-Caris, and G.J. Zweere, *J. Wat. Supply Res. Technol. Aqua* 1(69) (2020) 427–437.
2. T. Alelign and B. Petros, *Adv. Urol.* (2018) 3068365.

3. National Research Council (US) Safe Drinking Water Committee, *Drinking Water and Health: Volume 1*, National Academies Press, 1977, IV, Solid Particles in Suspension, https://www.ncbi.nlm.nih.gov/books/NBK234169/.

4. J. Zeng, G. Han, Q. Wu, and Y. Tang, *Int. J. Environ. Res. Pub. Health* 16(10) (2019) 1843, https://doi.org/10.3390/ijerph16101843.

5. IUPAC, *Compendium of Chemical Terminology*, 2nd ed. (The "Gold Book"), IUPAC, 1997, Online corrected version: (2006–) "Solubility", https://doi.org/10.1351/goldbook.S05740.

6. M. Clugston and R. Fleming, *Advanced Chemistry*, 1st ed., p. 108, Oxford Publishing, 2000.

7. Pharmacopeia of the United States of America, 32nd revision, and the National Formulary, 27th ed., pp. 1–12, 2009, https://search.worldcat.org/title/The-United-States-pharmacopeia.-32nd-revision-:-the-national-formulary.-27th-ed/oclc/822886535.

8. E. Rogers and I. Stovall, *Fundamentals of Chemistry: Solubility*, Department of Chemistry, University of Wisconsin, 2000, Archived from the original on 13 April 2015. Retrieved 22 April 2015.

9. S.E. Jørgensen (Ed.), Chapter 3 Precipitation, Coagulation and Flocculation, In *Studies in Environmental Science*, Vol. 5, pp. 25–37, Elsevier, 1979.

10. J. McGinty, N. Yazdanpanah, C. Price, J.H. ter Horst, and J. Sefcik, Chapter 1: Nucleation and Crystal Growth in Continuous Crystallization†, In *The Handbook of Continuous Crystallization*, Ed. N. Yazdanpanah and Z.K. Nagy, pp. 1–50, The Royal Society of Chemistry, 2020.

11. V.C. Gopalratnam, et al., *Environ. Prog.* 7 (1988) 84.

12. G. Hogarth, *Mini-Rev. Med. Chem.* 12 (2012) 1202–1215.

13. J.-Q. Jiang, *Sep. Purif. Rev.* 30 (2001) 127–141.

14. J. Bratby, *Coagulation and Flocculation in Water and Wastewater Treatment*, 2nd ed., IWA Publishing, 2006.

15. N.L. Nemerow (Ed.), *Industrial Waste Treatment, Chapter 6 – Removal of Colloidal Solids*, pp. 79–88, Butterworth-Heinemann, 2007.

16. H. Wei, B. Gao, J. Ren, A. Li, and H. Yang, *Wat. Res.* 143 (2018) 608–631.

17. J. Cohen and Z. Priel, *J. Chem. Phys.* 93 (1990) 9062.

18. V. Chandrasekhar, *Inorganic and Organometallic Polymers*, pp. 108–112, Springer, 2005.

19. B. Bolto and J. Gregory, *Wat. Res.* 41 (2007) 2301–2324.

20. A.K. Tolkou and A.I. Zouboulis, *Des. Wat. Treat.* 53 (2015) 3309–3318.

21. L. Lunevich, Aqueous Silica and Silica Polymerisation, In *Desalination – Challenges and Opportunities*, Ed. M.H. Davood Abadi Farahani, V. Vatanpour, and A.H. Taheri, 15 July 2020, https://doi.org/10.5772/intechopen.84824.

22. E. Katoueizadeh, M. Rasouli, and S.M. Zebarjad, *J. Mater. Res. Technol.* 9 (2020) 10157–10165.

23. A.I. Zouboulis and N.D. Tzoupanos, *J. Haz. Mat.* 162 (2009) 1379–1389.

24. M. Daifa, E. Shmoeli, and A.J. Domb, *Polym. Adv. Technol.* 30 (2019) 2636–2646.

25. J. Turunen, A. Karppinen, and R. Ihme, *SN Appl. Sci.* 1 (2019) 210.

26. V.A. Joshi and M.V. Nanoti, *Ind. J. Environ. Prot.* 19 (1999) 451–455.

27. K. Terada, *Encyclopedia of Separation Science*, Ed. I.D. Wilson, pp. 4394–4402, Academic Press, 2000.

28. A. Gupta and D. Yan, Chapter 14 – Solid – Liquid Separation – Thickening, In *Mineral Processing Design and Operations*, 2nd ed., pp. 471–506, Elsevier, 2016.

29. S. Pincetl, A Living City: Using Urban Metabolism Analysis to View Cities as Life Forms, In *Metropolitan Sustainability*, Ed. F. Zeman, Woodhead Publishing Series in Energy, pp. 3–25, Woodhead Publishing, 2012.

30. L.K. Wang, Y.-T. Hung, H.H. Lo, and C. Yapijakis, *Handbook of Industrial and Hazardous Wastes Treatment*, 2nd ed., CRC Press, 2004.

31. S. De Gisi and M. Notarnicola, Industrial Wastewater Treatment, In *Encyclopedia of Sustainable Technologies*, Ed. M.A. Abraham, pp. 23–42, Elsevier, 2017.

32. S. Judd and C. Judd (Eds.), *Chapter 2 – Fundamentals*, pp. 21–121. The MBR Book, Elsevier Science, 2006.

33. Y.-S. Chen and S.-S. Hsiau, *Powder Technol.* 192 (2009) 217–224.

34. F. Ruslim, H. Nirschl, W. Stahl, and P. Carvin, *Chem. Eng. Sci.* 62 (2007) 3951–3961.

35. D.B. Purchas, *Industrial Filtration of Liquids*, 2nd ed., Elsevier, 1971.

36. P. Ranjan and M. Prem, *Int. J. Curr. Microbiol. App. Sci.* 7 (2018) 637–645.

37. D.D. Ratnayaka, M.J. Brandt, and K. Michael Johnson, *Chapter 8: Water Supply*, Ed. D.D. Ratnayaka, M.J. Brandt, and K. Michael Johnson, 8th ed., pp. 315–350, Butterworth-Heinemann, 2009.

38. S. Garcia-Segura, M. Maesia, S.G. Eiband, J.V. de Melo, and C.A. Martínez-Huitle, *J. Electroanal. Chem.* 801 (2017) 267–299.

39. M.A. McDonald, H. Salami, P.R. Harris, C.E. Lagerman, X. Yang, A.S. Bommarius, M.A. Grover, and R.W. Rousseau, *React. Chem. Eng.* 6 (2021) 364–400.

40. M. Giulietti, M.M. Seckler, S. Derenzo, M.I. Ré, and E. Cekinski, *Chem. Eng.* 18 (2001).

41. E.A. López-Maldonado, M.T. Oropeza-Guzman, J.L. Jurado-Baizaval, and A. Ochoa-Terán, *J. Hazard. Mater.* 279 (2014) 1–10.

42. H.M. Abdulla, E.M. Kamal, A.H. Mohamed, and A.D. El-Bassuony, *Proceeding Fifth Science Environmental Conference*, Vol. VI, pp. 171–183, Springer, 2010.

43. S. Kocaoba and G. Akcin, *Talanta* 57(1) (2002) 23–30.

44. C.R. Ramakrishnaiah and B. Prathima, *Int. J. Eng. Res. Appl.* 2(2) (2012) 599–603.

45. F. Minas, B.S. Chandravanshi, and S. Leta, *Chem. Int.* 3(4) (2017) 291–305.

46. M.M. Brbooti, B.A. Abid, and N.M. Al-Shuwaiki, *Eng. Technol. J.* 29 (2011) 595–612.

47. A. Esmaeili, A. Mesdaghinia, and R. Vazirineja, *Am. J. Appl. Sci.* 2(10) (2009) 1471–1473.

48. Z.R. Guo, G. Zhang, J. Fang, and X. Dou, *J. Clean. Prod.* 14(1) (2006) 75–79.

49. T. Karidakis, S. Agatzini-Leonardou, and P. Neou-Syngouna, *Hydromet.* 76 (2005) 105–114.

50. D. Bhattacharyya, A. B. Jumawan Jr., and R.B. Grieves, *Sep. Sci. Technol.* 14 (1979) 441–452.

51. V. Yatirajam, U. Ahuja, and L.R. Kakkar, *Talanta* 23 (1976) 819–822.

52. W. Liu, B. Sun, D. Zhang, et al., *JOM* 69 (2017) 2358–2363.

53. Q.-Q. Lin, G.-H. Gu, H. Wang, C.-Q. Wang, Y.-C. Liu, R.-F. Zhu, J.-G. Fu, *Trans. Nonferr. Met. Soc. China* 26 (2016) 1118–1125.

54. K. Yan, Z. Liu, Z. Li, R. Yue, F. Guo, and Z. Xu, *Hydromet.* 186 (2019) 42–49.

55. L. Thanh and J.C. Liu, *Water Sci. Technol.: J. Int. Asso. Wat. Poll. Res.* 75 (2017) 2520–2526.

56. F. Tassel, J. Rubio, M. Misra, and B.C. Jena, *Min. Eng.* 10 (1997) 803–811.

57. H. Zhena, Q. Xu, Y. Hu, and J. Cheng, *Chem. Eng. J.* 209 (2012) 547–557.

58. F. Fu, R. Chen, and Y. Xiong, *Sep. Purif. Technol.* 52 (2006) 388–393.

59. M.E. Andrus, *Metal Finishing* 11 (2000) 20–23.

60. M.A. Barakat, *Arab. J. Chem.* 4 (2011) 361–377.

61. A. Pohl, *Water Air Soil Pollut.* 231 (2020) 503.

62. D. Polo-Cerón, *Bioinorg. Chem. App.* (2019) Article ID 3520837, https://doi.org/10.1155/2019/3520837.

63. M. Matlock, B. Howerton, K. Henke, and D. Atwood, *J. Haz. Mater.* 82 (2001). 55–63.

64. F. Xu, L. Xie, B. Tang, Q. Wang, and S. Jiang, *Chem. Eng. J.* 189–190 (2012) 283–287.

65. J. Singh and B.-K. Lee, *J. Environ. Manag. 161* (2015) 1–10.

66. H. Li, et al., *J. Clean. Prod.* 221 (2019) 89–97.
67. J. Lan, et al., *Chem. Eng. J.* 359 (2019) 1139–1149.
68. L. Wang, et al., *Environ. Sci. Pollut. Res.* 26 (2019) 29736–29747.
69. M.-X. Zhu, L. Lee, H.-H. Wang, and Z. Wang, *J. Haz. Mat.* 149 (2007) 735–741.
70. A.Z. Bouyakoub, B. Lartiges, R. Ouhib, S. Kacha, A. El Samrani, J. Ghanbaja, and B. Odile, *J. Haz. Mat.* 187 (2011) 264–273.
71. G. Zhang, X. Li, Y. Li, T. Wu, D. Sun, and F. Lu, *Desal.* 274 (2011) 255–261.
72. R.J. Morton and C.P. Straub, *J. Am. Wat. Works Assoc.* 48 (1956) 545–558.
73. W. Kim, Y.-J. Baek, K.-Y. Lee, D.-Y. Chung, and J.-K. Moon, *J. Nuc. Sci. Technol.* 53 (2016) 439–450.
74. A. Baeza, M. Fernández, M. Herranz, F. Legarda, C. Miró, and A. Salas, *J. Radioanal. Nuc. Chem.* 260 (2004) 321–326.
75. T. Gäfvert, C. Ellmark, and E. Holm, *J. Environ. Radioac.* 63 (2002) 105–115.
76. N.A. Oladoja, I.O. Raji, S.E. Olaseni, and T.D. Onimisi, *Chem. Eng. J.* 171 (2011) 941–950.
77. T. Muddemann, D. Haupt, M. Sievers, and U. Kunz, *Chem. Bio. Eng. Rev.* 6 (2019) 142–156.
78. K.E. Lee, N. Morad, T.T. Teng, and B.T. Poh, *APCBEE Procedia* 1 (2012) 59–65.
79. B. Merzouk, G. Bouchaib, K. Madani, C. Vial, and A. Sekki, *Desal.* 272 (2011) 246–253.
80. T. Kim, E.-B. Shin, and S. Kim, *Desal.* 161 (2004) 49–58.
81. E. Butler, Y.-T. Hung, R. Yu-Li Yeh, and M.S. Al Ahmad, *Water* 3 (2011) 495–525.
82. H. Patel and R.T. Vashi, *Global Nest J.* 15 (2013) 522–528.
83. L.F. Albuquerque, A.A. Salgueiro, J.L. Melo, and O. Chiavone-Filho, *Sep. Purif. Technol.* 104 (2013) 246–249.
84. Y.F. Wang, B.Y. Gao, Q.Y. Yue, Y. Wang, and Z.L. Yang, *Biores. Technol.* 113 (2012) 265–271.

7 Chromatographic techniques
Workhorse of analytical chemistry

7.1 WHAT IS ANALYTICAL CHEMISTRY?

Analytical chemistry is too broad and too active a discipline to be defined completely. It is often described as the area of chemistry used to characterize matter by answering the two fundamental questions of what is there (*qualitative*) and how much is there (*quantitative*). Thus, if we need to know if a particular species is present in a sample, then it is known as qualitative analysis, and if we want to know how much of the species is present in a sample, then it is known as quantitative analysis. In daily routine, both qualitative and quantitative measurements are carried out. Analytical chemistry can be understood to be the process of improving established methods to counter new challenges that arise due to the nature of samples [1]. An exceptional feature of analytical chemistry is the advance of new analytical methods/techniques to address scientific questions. This field can be defined as "the science of inventing and applying the concepts, principles, and . . . strategies for measuring the characteristics of chemical systems" [2]. If these new analytical methods are adopted by other scientists, they may serve as enabling technologies that can advance the work of an entire field. Charles N. Reilley said: "Analytical chemistry is what analytical chemists do". Izaak Maurits Kolthoff (1894–1993) is widely regarded as the father of modern analytical chemistry. His research transformed the ways by which scientists separate, identify, and quantify chemical substances and built the field upon solid theoretical principles and experimental techniques. Kolthoff had commented during an interview at Beckman Center (15 March 1984):

> In 1912 I went to a book sale and bought ten books for fifty cents. One of the books was by Ostwald The Scientific Foundations of Analytical Chemistry. Ostwald wrote at the beginning of that book that analytical chemists are the maidservants of other chemists. This made quite an impression on me, because I didn't want to be a maidservant.

This comment is sent as recorded on tape held by The Chemical Heritage Foundation, Philadelphia, and the quotation was provided by W. H. Brock.

Analytical chemistry is carried by obtaining answers for the basic questions of what, where, how much, and what form about a material sample. Analyte is the constituent of interest, while matrix is the constituent except for the analyte in a sample. The discipline has expanded beyond the bounds of just chemistry, and many have

DOI: 10.1201/9781003442516-7

advocated using the name analytical science to describe the field. Analytical chemists use the analytical approach for problem solving. An analytical approach involves the following steps: (i) identifying and defining of the problem, (ii) designing or planning the experimental procedure, (iii) carrying out experiments and collecting relevant data, (iv) analysing the data, and finally (v) providing a solution to the problem. In the planning stage, the most important step is the selection of the appropriate method. Upon selection of the method, sampling is done to obtain samples for further analysis. The samples obtained are prepared and pretreated prior to analysis, and the appropriate parameters need to be measured. The Valid Analytical Measurement (VAM) programme was set up in 1988 by the Department of Trade and Industry (DTI) to improve the quality of analytical measurements. This was replaced by the National Measurement System (NMS). According to this system [3], measurements should be made for a requirement using tested methods and calibrated instruments by a qualified and competent analyst. It is necessary that the measurements need to be checked with those in other laboratories using the same well-defined procedures. It is also necessary to carry out continuous assessment of the laboratory [3]. The validation should check the recovery and also the values obtained with an alternate protocol. It also involves replicate analyses and tests on blanks and the use of certified materials. Proficiency testing schemes between laboratories are also crucial. Method validation is crucial to establish reference methods. This is carried out using relevant performance indicators such as selectivity, specificity, accuracy, precision, linearity, range, limit of detection (LOD), limit of quantification (LOQ), ruggedness, and robustness.

(i) Specificity and selectivity are performance characteristics of analytical methods frequently used in analytical literature [4]. Specificity is the ability to assess the exact components in a mixture and gives a measure of the degree of interference by other substances in a sample on the analysis of a particular analyte. However, it does not require the identification of all components in the sample. According to the official guideline to be applied for method validation ICH Q2(R1), specificity is defined as the "ability to assess unequivocally the analyte in the presence of components which may be expected to be present". Specificity refers to the degree of interference by other substances also present in the sample while analysing the analyte. For analytical methods, specificity defines the identity of an analyte among a mixture of similar components in a sample, where the identity of the components is not important. The second approach known as matrix interference studies is applied to evaluate if the matrix may have an influence on the results. Quality control (QC) laboratory methods for identification tests, impurity tests, and assays rely upon specificity. According to IUPAC, "Specificity is the ultimate of Selectivity" [5, 6]. Selectivity is the ability to differentiate the components in a mixture and is defined as "The analytical method should be able to differentiate the analyte(s) of interest and IS [IS = internal standard] from endogenous components in the matrix or other components in the sample". Specificity indicates that the method can respond to only one analyte. In the context of analytical chemistry, selectivity is preferred as per IUPAC recommendations.

(ii) Types of errors: The main aim of carrying out an analysis is to have a correct measurement. However, what does this mean? It is important to realize that the meaning of the term correct is valid only if the analyst is aware of the true value. Nevertheless, in practice, this is not the case, and a measurement will always be associated with an uncertainty with respect to this value that is an inherent error. In order to achieve the correct value, the analysis is carried out in replicate. Error is the difference between the true and experimentally obtained values. However, if the true value is known, then analysis is not required. The main aim of an analyst is to minimize and quantify errors. There are three categories of errors, namely, gross error (human error), systematic error (bias), and random error. Gross error includes human errors while performing measurement and recording data. This can be spotted from values that are very different from the set of values. Data points that statistically fall outside the range of a data set are called outliers. Gross errors can be reduced by (i) careful recording of data and subsequent calculation and (ii) increasing the number of analysts. A systematic error can be described as a measurement that is always too high or always too low, and the magnitude of the deviation from the "true" value is constant. A systematic error is often difficult to identify. These errors can be divided into subgroups, namely, environmental errors, observational errors, and instrumental errors. Environmental errors arise in the measurement due to the effect of the external conditions (temperature, pressure, humidity, and external magnetic field). Observational errors are the errors due to an individual's bias or carelessness, or lack of proper setting of the apparatus. Measurement errors include wrong readings due to parallax errors. Instrumental errors arise due to faulty construction and calibration of the measuring instruments. Such errors arise due to the hysteresis of the equipment or due to friction. Zero error (positive or negative) is a very common type of error (especially in devices such as Vernier calipers and screw gauge). Sometimes the readings of the scale are erased off, resulting in errors. Instrumental errors can be due to an inherent constraint of devices, misuse of apparatus, and effect of loading. Instrumental systematic errors can result from drift noise, external interference, or improper calibration of the instrument. These errors can be identified by analysing signals of standards on a regular basis; e.g. baseline drift is a common problem in atomic absorption spectroscopy (AAS) analysis, and so it is a common practice to analyse a blank and a known standard after every 5 or 10 samples. Chemical systematic errors can occur in the wrong standards used for calibration or from the reagents used in various steps (e.g. yield in derivatization prior to the analysis of the derivative, or loss in extraction prior to analysis). Random errors are those errors that occur irregularly due to arbitrary and unpredictable variations in experimental conditions (e.g. unpredictable fluctuations in temperature, voltage supply, etc.). An absolute error is calculated by the difference between the measured value of a quantity and its actual value, while a relative error is the ratio of absolute errors to the measured value.

(iii) Precision and accuracy of a method are two important parameters. Precision of a method is the degree of agreement among individual test results when the procedure is applied repeatedly to multiple samples. Precision is related to reproducibility and repeatability. This can be measured by analysing a series of samples from multiple samplings of a homogeneous lot. Accuracy is the degree of closeness of measurements of a quantity to that quantity's true value.

(iv) Statistical tools: These are used to understand the analytical data. The field of statistics, where the interpretation of measurements plays a central role, prefers to use the terms bias and variability instead of accuracy and precision: bias is the amount of inaccuracy, and variability is the amount of imprecision. For the measured values, two criteria are used to compare these results, namely, the average value or mean (technically known as a measure of location) and the degree of spread (or dispersion). The average value or mean is computed by the ratio of sum of all the values to the number of measurements. Spread or dispersion is the difference between the highest and lowest values. A more useful measure that utilizes all the values is standard deviation (SD) (σ), which is defined as the measure of the dispersion of data from its mean. It measures the absolute variability of a distribution. The higher the dispersion or variability of data, the larger the standard deviation and the larger the magnitude of the deviation of value from the mean, and vice versa. It is important to observe that the SD value can never be negative. There are two types of SDs, namely, population SD and sample SD. The sample mean is the average value for a finite set of replicate measurements on a sample. It provides an estimate of the population mean for the sample using a specific measurement method. Mean is obtained by dividing the sum of all the data points by the number of data points and is calculated by the formula of $\left(\sum \times i\right)/n$. SD is the most basic statistic related to uncertainty. Suppose a chemist analysed one solution n times to estimate the concentration of a given substance. SD represents the variability in the data set of n results and is calculated by $\mathrm{SD} = [\sum\left(x - x_{\mathrm{avg}}\right)^2 / \left(n - 1\right)]^{0.5}$. In statistics, SD (σ) is a measure of the amount of variation or dispersion of a set of values. A low SD indicates that the values tend to be close to the mean (also called the expected value) of the set, while a high SD indicates that the values are spread out over a wider range. The square of SD (σ) is a very important statistical quantity known as variance. Precision is calculated as relative SD (% RSD), using the relation $\%\mathrm{RSD} = \left(\mathrm{SD}*100\right)/100$. Thus in chromatography, all the considerations of analytical chemistry are also applicable.

7.2   WHAT IS CHROMATOGRAPHY?

The historical perspective of modern chromatography is as early as the late 19th and early 20th centuries [7]. It originated from the works of David T. Day, a distinguished American geologist, and Mikhail Tswett, a Russian botanist. Day developed procedures for crude petroleum fractionation using Fuller's earth [8], while Tswett used a chalk-filled column packed for separation of pigments of leaves [9]. Tswett was the one to interpret the process and termed it as chromatography (meaning "to write with odours" – literally translated from its Greek roots *chroma* and *graphein*), and so he is known as the father of chromatography. The revolution of chromatography was in the 1940s due to the development of column partition chromatography by A.J.P. Martin and R.L.M. Synge [10], for which they were awarded the Nobel Prize in 1952. In partition chromatography, the distribution of the solute occurs between two liquid phases, of which one of them is within a solid support. Thus, techniques such as paper and thin layer chromatography were developed in the early 1940s [11–14]. The 1950s was dominated by the development of ion exchange chromatography in the early years [15], while in the later part of the decade, the concept of gel filtration chromatography was introduced, paving the way for the development of gel permeation chromatography (GPC) in early 1960s [16]. The use of ion exchangers became quite common in the 1930s [17, 18], but their application for chromatographic determination of organic compounds was reported in 1958 [19]. This system was a precursor of high-pressure liquid chromatography (HPLC) that incorporated automatic pumping, efficient ion-exchange chromatography (IEC) columns, and continuous odour detection. Prior to the development of the first HPLC systems, gas chromatography (GC) gave an idea of the possible applications and ease of automation and sensitivity of HPLC. The early 1960s saw the automation of liquid chromatography (LC) and GPC [20–22]. The use of smaller particles in well-packed beds could increase both separation speed and efficiency, but high pressure is needed to pump the mobile phase through the column. Later, HPLC was developed [23], and by 1970, HPLC was marketed by Waters Associates and Du Pont. Horváth, Huber, and Kirkland can be considered the "fathers" of HPLC [24].

Chromatography refers to the separation techniques based on the partitioning or distribution of a sample (solute) between a moving or mobile phase (MP) and a fixed or stationary phase (SP) [25–27]. The process can be considered to be a combination of a series of equilibrations between the MP and SP. This phenomena can be mathematically expressed by the partition (K) or distribution (D) coefficient (ratio of concentrations of solute in SP to MP). Separation arises due to the differential affinity of the solute for the two phases. Therefore, if it has affinity for the MP, then it will move rapidly through the column and be detected earlier than the one that shows more affinity for the SP. Chromatography is based on a physical equilibrium that results when a solute is transferred between the MP and SP. By definition, chromatography is a separation technique in which a sample is equilibrated between a MP and a SP. The two phases are chosen such that the degree of distribution of various components of the sample between two phases is variable. Normally, the SP is a solid or liquid entrapped in a solid, while the MP can be a liquid, gas, or supercritical fluid.

7.3 CLASSIFICATION OF CHROMATOGRAPHY

The chromatography technique is classified into different categories depending on the technique involved or physicochemical principles involved.

The first type of classification is based on the combination of different parameters, namely, solute–SP interactions, chromatographic bed shape, physical state of the MP, and molecular characteristics. Therefore, the techniques can be classified into column chromatography, paper chromatography, thin-layer chromatography (TLC), gas chromatography (GC), HPLC, IEC, gel filtration chromatography, and supercritical fluid chromatography (SFC). On the basis of solute interactions with the SP, the techniques can be classified into adsorption, partition, ion exchange, molecular exclusion, and affinity chromatography. Partition chromatography utilizes a mobile liquid or gaseous phase entrapped within a solid support, while the mobile liquid sorbed on the surface of the SP acts as the SP. In IEC, a stationary solid phase covalently attaches anions or cations onto it. Thus, solute ions of the opposite charge in the mobile liquid phase are attracted to the resin by electrostatic forces. The difference in the extent of interaction leads to the separation of inorganic ions. In molecular exclusion chromatography, the liquid or gaseous phase passes through a porous gel that separates the molecules according to its size. Affinity chromatography utilizes the specific interaction between one kind of solute molecule and a second molecule that is immobilized on a SP. Thus, the above techniques can be classified based on the bed shape, namely, planar and column chromatography. In planar chromatography, the SP is supported on a flat plate or in the pores of a paper. The MP moves through the SP by capillary action or under the influence of gravity. Paper chromatography and TLC techniques are planar in configuration. Paper chromatography, introduced in 1944, has a liquid SP and MP (partition chromatography). Paper serves as a support for the liquid SP, and the sample solution is applied as a spot or streak one half inch or more from the edge of a filter paper strip. The paper with the dried spot is dangled in a closed container with the MP. The solvent, while moving up the paper length, will take different components to different distances through capillary action. The SP in this chromatography is the water within the pores of the paper, making it a partition chromatography. 2D technique is used for complex mixtures. In this, the sample spot is first developed in one direction using one solvent and then dried. After this, the paper is turned to right angles and developed using the second solvent of different polarity. Paper chromatography can be made more selective by using ion exchange groups incorporated within the cellulose structure. Paper chromatography and TLC are used to characterize the components based on the relative mobility values, which depend on the quality of the SP and its thickness, humidity, temperature, etc. TLC was first described in 1938 and became popular as it replaced paper chromatography due to its speed, sensitivity, and reproducibility. The resolution in TLC is higher than that of paper chromatography due to the presence of smaller and regular particles in the plate compared to paper fibres. TLC has many advantages such as high sample throughput, low cost, minimum sample preparation, and retrieval of plates and finds applications in various fields. The most commonly used sorbents are silica gel, alumina, diatomaceous earth, and cellulose. The separation using silica gel, aluminium oxide, and magnesium silicate as sorbent is based on adsorption, while that with cellulose is a partition mechanism. Small and uniform particles of silica with polar or non-polar groups are used in high-performance thin-layer chromatography (HPTLC). Sorption TLC involves the activation of sorbents by drying at a specified temperature for a specified time. The initial protocol of TLC matches with that of paper chromatography. Once the zones are separated, the detection is carried out using an appropriate technique (either in situ or by scrapping of a band from the plate, eluting the component, and determining) like derivatization (char using sulphuric acid or produce a coloured complex using iodine), fluorescence, autoradiography, biological assay (measurement inhibition of cholinesterase activity by organophosphate pesticides), etc. Different factors such as nature of the compounds, type of SP, solvent strength (the higher the strength, the greater the R_f value), type of developing chamber, vapour phase conditions, and development mode affect separations. The other classification of chromatography based on the column bed is column chromatography. In this, the SP is held in a narrow tube and the MP passes through the column, taking along with it the more soluble solute faster than the more firmly held solute. Column liquid chromatography, or known popularly as column chromatography, is a very well used separation technique involving the differential migration of solutes within a closed tube containing a packed SP. The liquid MP passes through a SP (solid or liquid impregnated into an inert solid) at low pressure conditions or under gravity. The most common technique for wet packing of a column involves the pouring of sorbent slurry into the column and draining of excess solvent once the sorbent is uniformly packed into the column. The packing may require repeated pouring of slurry and draining of excess solvent. The sample dissolved in a minimum volume of the MP is poured at the top of the column, while the eluent is passed at a constant rate using a peristaltic pump. The MP composition can be kept constant throughout the process (isocractic) or varying (gradient). The eluate coming from the column passes to a detector, wherein the individual components are analysed.

Based on the physical state of the MP, the column chromatographic techniques can be classified into LC, GC, and SFC techniques. In gas–liquid chromatography, the SP is a non-volatile liquid coated onto a porous support or the walls of a capillary tube, while the MP is an inert carrier gas (helium). Separation occurs due to differences in vapour pressure, and the solute with higher vapour pressures pass through the column faster. GC can be either partition or sorption based on the column material used. In HPLC (or more commonly LC), separation occurs due to adsorption on a column through which the MP is passed through under pressure. The column material can lead to separation by different mechanisms such as partition, ion exchange, or molecular permeation. IEC is a liquid–solid chromatography used to analyse ions as the SP is functionalized appropriately. In gel filtration chromatography, the SP is a gel, and separation is achieved using a HPLC instrument by the sieving mechanism. SFC uses carbon dioxide supercritical fluid at conditions above the critical pressure $\left(P_c\right)$ and critical temperature $\left(T_c\right)$. Carbon dioxide is not suited for polar and high-molecular-weight compounds, and therefore, small amounts of a polar solvent (methanol) is added. This results in enhanced solute solubility and leads to better peak shapes and efficient separation. There are other supercritical fluids like nitrous oxide, trifluoromethane, sulphur hexafluoride, pentane, and ammonia which find applications in food industries. SFC has properties transitional to both LC and GC but with unique features of fast analysis, better resolution, and high selectivity range by varying the pressure, temperature, MP composition, SP, etc. SFC can also be used for the analysis of non-volatile, thermally labile compounds (unlike GC). SFC can be performed using either packed columns (with small size, porous, high surface area, hydrated silica particles like in HPLC) or capillaries (coated with polysiloxane film whose polarity is varied using different functional groups). The instrument of packed-column SFC is similar to that of HPLC, but a back pressure regulator is used to control the outlet pressure of the system. The separations can be carried out in an enclosed bed (column) or in an open bed (thin-layer plate coated with an appropriate SP). Liquid–solid adsorption chromatography uses a column (packed with an adsorbent like alumina or silica gel), and the sample is eluted with a suitable solvent by gravity flow through the column. In column separations, MP flow is achieved by a pressure drop along the column, while in open-bed chromatography the flow is due to capillary wetting. In open-bed chromatography, the SP is open and no column is used. This is known as planar chromatography.

The second classification is based on the retention mode on the SP. The retention of the solutes on the SP can be through any of the following mechanisms, namely, sorption, exclusion, or ion exchange. Sorption is a general term used for both adsorption and partition of the solutes onto the SP. Chromatographic separation can be based on only adsorption or partition, leading to adsorption or partition chromatography. Exclusion refers to the separation technique based on sample size. This is the sieving mechanism that operates in gel permeation chromatography or size-exclusion chromatography. Ion exchange refers to the exchange of ions in a solution with those on the SP. Generally, the solutes are inorganic cations or anions. But organic ionic species can also be separated.

Chromatographic techniques can be classified based on the mode of sample introduction into the column as frontal chromatography, displacement chromatography, and elution chromatography. In frontal chromatography, the sample is introduced into the bed continuously, and it behaves like a MP. The sample components are selectively separated as fronts rather than as bands, with the least retained component emerging out of the column. Therefore, complete recovery of the pure sample is difficult. In displacement chromatography, the MP is much more strongly retained by the SP compared to the sample, leading to dislodgment of the sample by the MP. The main advantage of this technique is that greater loads of sample can be applied and can be useful for preparative- or production-scale samples. However, the main hitch is to find the correct displacing reagent. This is useful in the bioseparation of proteins. Elution chromatography involves the separation of components into bands due to the migration of sample components at a pace slower than the MP at pre-set experimental conditions. This is exploited solely for analytical determinations. Elution separations can be non-linear or linear (as determined by the peak under the conditions of separations). Non-linear elution is usually observed when high sample concentrations are used (preparative separations), and so Gaussian bands are not obtained with tailing or leading of peaks. Since distribution constant K depends on the sample size, it becomes difficult to identify compounds based on their migration rates. Therefore, most of the separations (analytical/preparative) are carried out under linear isotherm conditions, wherein the distribution constant K is independent of the sample concentration.

LC is a column separation method used for separating and analysing compounds based on differences in their interaction with a SP. The interactions can be adsorption, partition, ion exchange, molecular exclusion, or affinity. There are many different ways to use LC for separation purpose, and the choice depends on the physicochemical characteristics of the molecule of interest. The solid phases are highly porous, chemically inert supports functionalized with various chemical groups that control the interactions with the molecules to be separated. The different commonly used modes of separation are based on specific binding interactions (affinity chromatography), charge (IEC), size (size exclusion chromatography/gel filtration chromatography), and hydrophobic surface area (normal/reverse phase chromatography; NPC/RPC). In this chapter, further discussion will be restricted to the chromatographic techniques normally employed for the determination of inorganic ions. The most commonly used technique is the ion chromatography, while reverse-phase HPLC also finds applications in these analyses.

7.4 BASIC PRINCIPLES OF LIQUID CHROMATOGRAPHY

The basic principle is based on the movement of analytes through the column which is governed by their partition coefficients between the two phases [28]. The concentrations involved in chromatographic analysis are quite low, and the problems of saturation leading to column overloading and subsequent problems of loss of column efficiency and poor resolution do not arise [29]. The process of chromatography produces peaks of analyte components due to separation into discrete bands. However, optimization of a successful separation requires that researchers are aware of the various aspects of chromatography apart from its instrumentation. The chromatogram is a plot of detector response produced due to the entry of the analyte into the detector, as a function of elution time. The peaks may be symmetrical or asymmetrical in nature based on the entire separation process. When the volume of the eluent is considered instead of the time of elution, it is possible to obtain void volume, which is defined as the time taken by one volume of the MP [30]. If the solute is partitioned between the two phases, then the volume of the eluent that passes through the column before the solute elutes out is known as retention volume V_T, and it is the sum of the void volume $\left(Vm\right)$ and the volume fraction due to the partitioning between the two phases $\left(Vm*Kx\right)$. The time taken for the solute to reach out of the column is known as the retention time. It is quite understandable that different components will have different retention time values due to their differences in partitioning efficiency between the two phases. The volume of the MP required to elute a component from the SP and out of a chromatography column is called the retention volume (V_R), and the time associated is t_R [31]. The free space between the SP particles packed in the column is known as void volume V_0. All molecules spend the same total amount of time in the MP, and this time is called the column dead time or holdup time, t_0. This time is measured by injecting a completely unretained compound and measuring the time required for the peak (measured at peak height) to reach the detector. The retention time, t, is the time from when the sample is introduced into the column to when the detector senses the maximum of the retained peak. This value is greater than the column holdup time by the amount of time that the compound has spent in the SP and is called the adjusted retention time (t_R'). If the analytes do not partition between the two phases, then the retention time will be the same and is denoted as t_M. The adjusted retention time t_R' is the time for which the analyte is retained on the SP, and it is given by $t_R' = t_R - t_M$. The retention factor or the capacity factor $\left(k\right)$ explains the migration rate of an analyte in a given column. It is represented as

$k' = t_R - t_M / t_M = t'_R / t_M$. The retention factor has no units, and ideally it should be between 1 and 10. If the value is less than 1, it indicates very fast elution, thus making it difficult to achieve precise determination; a value greater than 10 is indicative of very slow separation. The separation of two molecules (A and B) on a column is expressed by the selectivity factor α, which is given as the ratio of capacity factors of the two solutes.

The peak broadening that occurs quite commonly is a general term used to describe the overall dispersion or widening of a sample peak as it passes through a separation system. It occurs as a result of several factors such as diffusion processes and solute distribution between phases. The width of the peak is quite important to understand various concepts of chromatography. Resolution (R_s) may be expressed by Eq. 7.1, wherein $\Delta t =$ difference between retention times of the two solutes with peak widths at baseline of w1 and w2, respectively [32–36]. Resolution is greatly affected by column efficiency, selectivity, and capacity and can be mathematically expressed by Eq. 7.2, wherein the first, second, and third terms denote the column efficiency, selectivity, and capacity, respectively. In case of poor resolution, the column efficiency should be the first parameter that has to be evaluated. Separation on an efficient column results in narrow and less spread peaks in the chromatogram.

$$R_s = \frac{2\Delta t}{\left(w_2 + w_1\right)} \tag{7.1}$$

$$R_s = \frac{\sqrt{N}}{4} * \left(\frac{a-1}{a}\right) * \left(\frac{k'}{k'+1}\right) \tag{7.2}$$

When the partition coefficient (K) of a solute [K = (solute concentration in SP)/(solute concentration in MP)] is the same for all solute concentrations, the elution will ideally follow a symmetrical Gaussian peak. If the elution does not follow this trend, the chromatographic peak will be asymmetric in shape. The Gaussian nature of the chromatographic peak is understood using the Plate theory. The separation factor, α, is a measure of the resolution between two peaks and is the ratio of the adjusted retention time values of the two solutes. It is dependent on various factors such as temperature, SP, and MP.

Column efficiency can be explained by two approaches, namely, plate theory proposed by Martin and Synge in 1941, and rate theory, proposed by van Deemter in 1956 [37,38]. Both rate theory and plate theory are important in chromatographic analysis [39,40]. The major difference between the two theories is that rate theory describes the efficiency of separation by comparing the rates of analytes eluted through the column, whereas plate theory describes the process by the determination of the number of hypothetical plates (where distribution between two phases occur) in the column.

According to the plate theory, a chromatographic column of length L is viewed to be made up of a number (N) of series of narrow, discrete sections called theoretical plates of height equivalent H, which can be expressed by Eq. 7.3. The schematic representation of the plate theory is given in Figure 7.1. The figure shows that at each

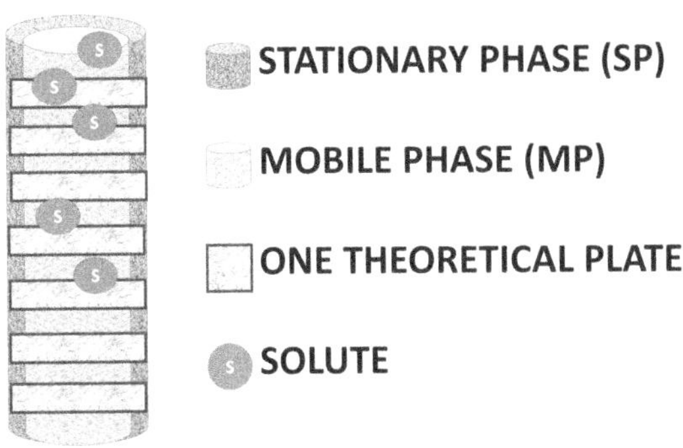

FIGURE 7.1   Schematic representation of plate theory of chromatography

plate, there is an equilibrium. In this, the analyte is distributed between the MP and SP. The movement of the analyte and MP is seen as a series of transfers from one theoretical plate to another. The correlation between the height equivalent of a theoretical plate to the column length (L) and number of plates (N) is given by Eq. 7.3:

$$HETP = \frac{L}{N} \tag{7.3}$$

Thus, it can be easily understood that column efficiency is higher if there are more numbers of theoretical plates due to the increased number of equilibrations. Also, the higher number of theoretical plates suggests that the height equivalent of a theoretical plate, i.e. height of each plate H, is smaller. Thus, smaller heights promote better equilibration between the SP and MP. Thus, columns with low plate numbers are less efficient than columns with higher plate counts. The peak parameters can be used for the calculation of N (number of theoretical plates). For a simple Gaussian peak, N can be related to the retention time, as given in Eq. 7.3a, wherein t_R is the retention time and σ is the SD. The different methods like the USP method and the half-peak-height method are used for the calculation of N using the peak features. Both the methods first involve the determination of peak width, which is the distance between points where lines tangent to the peak's left and right inflection points intersect the baseline. The calculations can be carried out assuming the chromatographic peak to be a Gaussian distribution (normal distribution), as shown in Figure 7.2 (a). The peak has a height of H and width of w, and the width at half peak height ($H_{0.5}$) is $w_{0.5}$. The SD of the peak is given by σ.

$$N = \left(\frac{t_R}{\sigma}\right)^2 \tag{7.3a}$$

H is given in terms of SD and also variance in Eqs. 7.3b and 7.3c, respectively.

$$N = \left(\frac{L}{\sigma}\right)^2 \tag{7.3b}$$

$$H = L * \left(\frac{\sigma}{L}\right)^2 = \frac{\sigma^2}{L} \tag{7.3c}$$

Since variances $\left(\sigma^2\right)$ are additive in nature, it can be safely understood that the total plate height observed for a system $\left(H_{tot}\right)$ is a sum of variances of plate height for each individual process, leading to an increase in peak band-broadening.

The *US Pharmacopeia* method of calculation of N involves the use of a simple relation given in Eq. 7.3d, wherein t_R is the retention time and w is the peak width [30,31]. This calculation results in small N values when peaks overlap or there is distortion in peak shape or the peak has multiple inflection points. The second method is the half-peak-height method, which involves the use of width at half the peak height $(w_{1/2})$. This is the most extensively used method and is also used by the *German Pharmacopeia* (DAB), *British Pharmacopoeia* (BP), and *European Pharmacopoeia* (EP). This involves the calculation of N by Eq. 7.3e.

$$N = 16 * \left(\frac{t_R}{w}\right)^2 \tag{7.3d}$$

$$N = 5.5 * \left(\frac{t_R}{w_{\frac{1}{2}}}\right)^2 \tag{7.3e}$$

Symmetry factor (S, also called "tailing factor") is a coefficient that shows the degree of peak symmetry. It is represented in Eq. 7.3f, wherein $w_{0.05}$ is the width at the height of 1/20th peak height. The value of f can be calculated from the peak shown in Figure 7.4B. The values of S indicates the nature of the peak; $S > 1, = 1$, and < 1 refer to the tailing peak, peak with Gaussian distribution (symmetry), and leading peak, respectively.

$$S = \frac{w_{0.05}}{f} \tag{7.3f}$$

The peak broadening due to the contribution from the solute movement through the chromatography column can be explained by rate theory. This is represented schematically in Figure 7.2.

The first factor is eddy diffusion, which occurs as the analyte molecules adopt different routes of travel across the column. These routes have different widths and lengths across the column. If the column is packed with very large and inhomogeneous particles, then eddy diffusion becomes a major factor responsible for peak broadening. In a packed column, the particle size is directly related to the height of the plate, as given by Eq. 7.4, wherein d_p is the diameter of a particle. Therefore, increased particle size leads to an increase in eddy diffusion, and so in open tubular columns, eddy diffusion is not a major concern. In LC, eddy diffusion is responsible for a major part of the peak or band broadening in the column. Since eddy diffusion is a combination of diffusion and convection, the term eddy dispersion might be

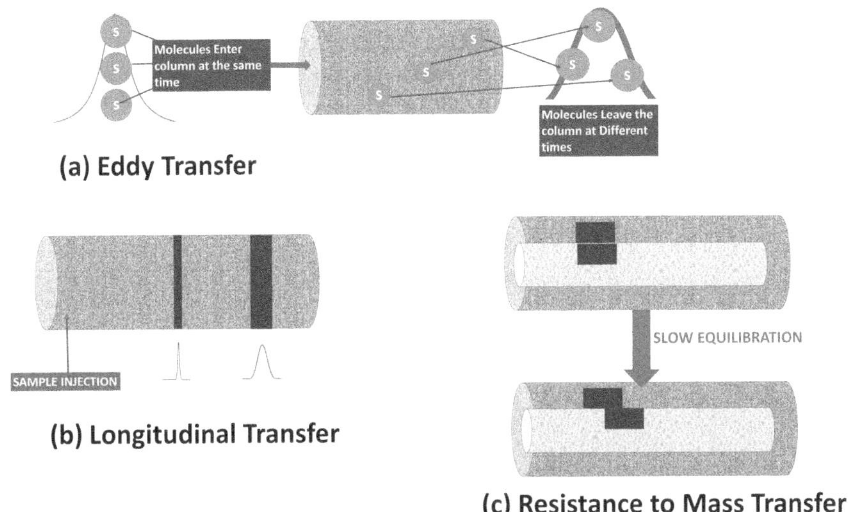

FIGURE 7.2 Schematic representation of rate theory of chromatography

more appropriate. The contributions to eddy dispersion arises from internal diameter, length, and packing efficiency of the column and also the size and homogeneity of particles.

$$H = C_e * d_p \tag{7.4}$$

The second factor that contributes to peak broadening is longitudinal diffusion. This is due to the fact that diffusion occurs across a concentration gradient from the concentrated region in the centre to a dilute one at boundary. The contribution of longitudinal band broadening is proportional to the diffusion constant. Since the diffusion in gases is about 104 times higher than that in liquids, longitudinal broadening is more prominent in GC than in HPLC. At high velocity of the MP, the analyte does not get sufficient time to be retained onto the column, and so the effect of longitudinal diffusion is reduced.

The third factor that contributes to peak broadening is the resistance to mass transfer. For every equilibrium between the two phases, the time involved tends to vary if the solute prefers one particular phase, leading to peak broadening. Therefore, this broadening is due to the analyte diffusion and convection between the two phases. The resistance to mass transfer is inversely relational to the diffusion constants in either phase. In the MP, there is an additional link to eddy diffusion. Thus, the total contribution to the plate height can be described by band broadening taking place in the MP, stagnant MP [41], and SP. Different mathematical equations are given to understand the relationship in the different phases.

(i) Resistance to mass transfer in the MP, or MP mass transfer is a process of peak broadening caused by the presence of different flow profiles within channels or between particles of the support in the column.

(ii) Resistance to mass transfer in the stagnant MP: This is common in a packed column where MP can be stagnant within the pores of SP and solutes diffuse into and out of it. This is a major source of band dispersion [32], which is affected by particle shape and size distribution and by the hydrodynamics of the stagnant boundary layer and packing density, which presents diffusional mass transfer resistance [36]. This affects the value of H as shown below.

(a) In an open tubular column, $H = \dfrac{C_m * d_c^2 * u}{D_m}$.

(b) In a packed column, $H = \dfrac{C_m * d_c^2 * u}{D_m}$, where $H = \dfrac{C_m * d_c^2 * u}{D_m}$.

(c) Resistance to mass transfer in the SP: It takes the solute molecules some time to reach the SP, to enter it, to remain there for a duration that is governed by statistics, and to leave the SP and enter the MP again.

$$C_m = \frac{1 + 6k + 11k^2}{96 * (1 + k)^2}$$

The van Deemter equation combines the different contributions in a simplified equation, wherein A is eddy diffusion, B is the longitudinal diffusion in the MP, and C is the resistance to mass transfer in the SP (in GC) or in both the phases (LC).

$$H = A + \frac{B}{u} + C * u$$

The contribution of a stagnant MP is valid only with large particles and is, therefore, usually neglected in the van Deemter equation for high-performance systems. The contribution of resistance to mass transfer in the MP can be neglected in GC, but not in HPLC.

The assumptions made in the classical derivation of the van Deemter height equivalent to a theoretical plate (HETP) equation show a mathematical expression between the Gaussian distribution curves of the solute concentrations in the eluent provided by the plate theory [37–40] and by the rate theory applied to a continuous column. In the rate theory, the axial dispersion coefficient $_{DL}$ (m^2/s) is applicable only to the MP of volume fraction F_1 as there is no axial dispersion in the SP of volume fraction $(1-F_1)$. The linearly flowing liquid phase moves the solute between itself and the SP. According to plate theory, it is assumed that the volume of eluent required to elute an analyte through one plate is lower than the elution volume (volume of the eluent that elutes from the column before a particular analyte comes out with the eluent). Similarly, rate theory considers the elution distance of an analyte to be greater than the sum of the distance in the mixing stage and the height of transfer. Another assumption is the effect of injection volume on the peak features, and this is applicable to GC. The fundamental assumptions of the van Deemter HETP equation were formulated in 1956. But it is seen that there are serious errors in the interpretations using the original equation. In this work, it was seen that there was an excellent fit of the equation for small molecules, but there are instances as to why this model becomes inadequate. Longitudinal diffusion takes place not only in the interparticle MP but also in the pores (within/surface). The overall eddy dispersion term is due not only to constant flow-controlled eddies but also to asymptotic transchannel and short-range interchannel eddy dispersion. The mass transfer resistance in the MP cannot be represented by a linear liquid driving force model as this approach leads to a retention-dependent HETP term. The mass transfer resistance in the SP cannot be accurately represented by a simplified version of diffusion across a spherical particle [41]. The accurate and correct expression was derived earlier.

Therefore, in a nutshell, it will prove to be useful to understand the differences between the two theories. Rate theory is a concept in chemistry that describes the process of peak dispersion, and it provides an equation to calculate the variance per unit length of the column. This theory is very useful in column chromatography. It describes the features of chromatographic separation by comparing the rate of the analyte that elutes through the column. So it provides a more realistic description of the processes that work inside the column. Plate theory is a concept that describes separation in the form of theoretical plates, which is the site where the equilibrations between the SP and MP occur. Thus, it provides a hypothetical description of the processes that work inside a column.

From the Van Deemter plot, the flow rate corresponding to minimum plate height and subsequently maximum column efficiency can be obtained. Values higher than this will lead to peak broadening. The temperature of separation will affect the longitudinal diffusion and the mass transfer, and an increase in temperature enhances the rate of movement of the solute between the two phases, resulting in faster elution and narrow peaks on the chromatogram.

Column selectivity also affects the resolution in addition to efficiency. Column selectivity is defined by the relative separation α between the two peaks and is represented mathematically as below, wherein t_{R1}, t_{R2}, and t_0 are the retention times of solutes 1, 2, and retained components, respectively, and K_1 and K_2 are the distribution coefficients of solutes 1 and 2, respectively. For a good resolution, selectivity is more important than efficiency as resolution is directly related to selectivity but is quadratically related to efficiency; thus, a fourfold increase in N is needed to double R_s.

$$\alpha = \frac{t_{R2} - t_0}{t_{R1} - t_0} = \frac{K_2}{K_1} \tag{7.4}$$

Band broadening can occur in the injector, connecting tubings (the parabolic flow profile caused by friction at the walls), inside the detector, and sometimes due to slow electronics in the detector or the data system.

Column capacity or retention factor k' is the time a solute spends in/on the SP relative to the MP. The low values of k' signify less retention, and components get eluted close to the solvent front, leading to poor separations, and this problem normally arises due to overuse/misuse of the column (loss of functional groups). The higher values of k' result in improved separation but can result in broad peaks and longer analysis time. For a good separation of practical importance, the k' values should be within the range of 1–15.

Quantitation is the final step in chromatography once well-resolved peaks are obtained. The peaks are identified by comparing V_R / t_R (or preferably the adjusted retention times) to that of standards separated under identical conditions. But there is always a danger of different compounds having identical retention times, and thus, this is countered by using different techniques like spiking of samples with known compounds and comparing the chromatograms of original and spiked samples to confirm the peak identity. In order to calculate the concentration of the solute using peak area or peak height, the quantification can be carried out using either external standard or internal standard methods. In the external standard method, a series of standards of known analyte concentration is analysed. The peak height or area is plotted as a function of analyte concentration, giving a curve known as calibration curve. The sample is then analysed under identical condition, and using its peak height or area, the concentration can be computed from the calibration curve. In this technique, irreproducible injections can cause errors in analysis. Therefore, automatic injectors and sampling valves can be used instead of manual injection to reduce these errors. If there are no suitable external standards available, then at times, the detector response factors may be used to allow calibration with one compound. Here, the ratio of concentrations of the unknown to known is equated to the response of the two, thus allowing easy calculation of the unknown concentration. In the internal standard method, an internal standard (compounds with sufficient similarity to the target analytes but not present in the sample) is added to the sample prior to analysis. Internal standard compounds must also be readily separated from any of the compounds in the sample. It is better to add internal standard compounds at the beginning itself, thus making the procedure quite easy to follow. The concentration of the internal standard should be in the same range as that of the sample. The area of the internal standard is used to normalize the areas of all other sample peaks, thus eliminating the effect of differences in injection volumes or dilutions.

7.5 INSTRUMENTATION OF LC

The basic components of a LC system are schematically represented in Figure 7.3. The process involves sample injection into a column packed with a SP, wherein the individual sample components traverse through the column by a liquid (MP). The components of the sample are separated from one another on the column due to chemical and/or physical interactions between solute components and the SP. The separated components are collected at the column exit and identified using a detector.

High-performance columns filled with packing materials of very small particle size (diameter: 3–10 microns) are used, making it essential to force the eluent using a high-pressure pump (unlike in conventional column chromatography where gravity plays a major role). The pump can be single-piston or dual-piston, which ensures a pulse-free flow of the eluent (using pulse dampeners with a single-piston pump and electronic circuitry with a dual-piston pump). The sample is injected into the system at atmospheric pressure via a three-way valve (volume of 10 μL–100 μL), of which two ports are connected to the sample. The valve is then switched to inject mode to

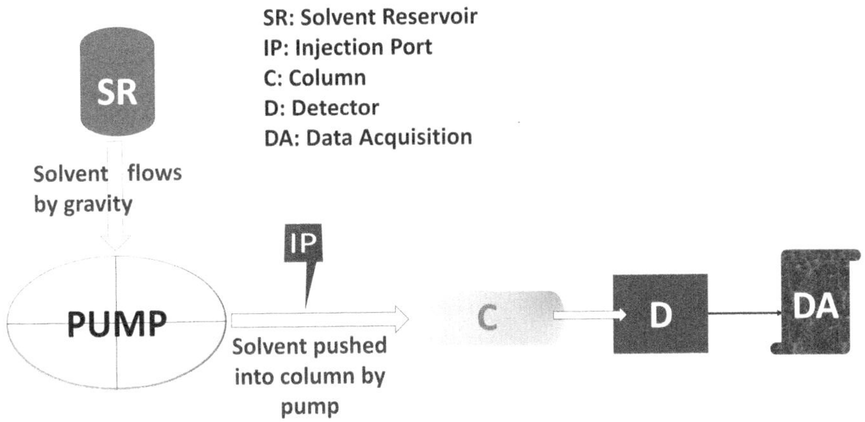

FIGURE 7.3 Schematic representation of instrumentation of LC (drawn with reference to https://commons.wikimedia.org/wiki/File:HPLC_apparatus.svg)

allow the sample to mix with the MP. After switching the injection valve, the sample is transported to the separator by the MP. The column is very important, and the choice of a suitable SP and the use of appropriate experimental conditions become very critical in deciding the quality of analysis. HPLC can be divided most commonly into RPC, NPC, and ion chromatography. The method of selection is based on the nature of the solutes to be separated.

7.5.1 Normal and Reverse Phase Chromatography

NPC was invented first, and hence it is termed normal. In this, a highly polar material (silica) is used as the SP, while non-polar solvents are used as the MP [42]. Alumina is favoured for the separation of unsaturated bonds. Moderately polar-bonded phases $\left(\text{egcyanopropyl-}(CH_2)3\text{–CN}, \quad \text{diol–}(CH_2)3\text{–O–}CH_2\text{–CHOH–}CH_2\text{–OH},\right.$ $\left.\text{aminopropyl–}(CH_2)3\text{–NH}_2\right)$, chemically bonded on a silica gel support can also be used in the NPC mode. This technique was earlier known as adsorption or liquid–solid chromatography as the retention on inorganic sorbents is due to interactions of the polar surface groups with the analytes. MP is a mixture of two or more organic solvents of different polarities. The polarity of SPs in NPC is stronger in comparison to that of the MP. Sample retention increases on more polar SPs and in less polar MPs. The strength of interactions with analytes increases in the order of cyanopropyl<diol<aminopropyl<< silica ≈ alumina SPs. The silanol groups in silica gel strongly attract basic analytes, while the presence of aminopropyl groups on silica makes it useful for the separation of acidic solutes (alcohols, esters, ethers, and ketones). Monolayer sorption is the main mechanism of retention of solutes on the SP by displacing the solvent molecules already sorbed onto the surface. This is governed by the sorption efficiency and also the solvent strength of the MP. In addition to this, liquid–liquid partition may also occur. It is essential to use dehydrated solvents to achieve excellent separations in NPC. The elution times in NPC is the least for alkanes and the highest for carboxylic acids (alkanes < alkenes < aromatic hydrocarbons ≈ chloroalkanes< sulphides < ethers < ketones ≈ aldehydes ≈ esters < alcohols < amides << phenols, amines, and carboxylic acids). The retention also decreases as the size of the alkyl group in the solute increases. NPC has several practical advantages like its applicability for the separation of positional isomers (due to different shapes such as rigid planar, rod-like, or flexible chain structure) and reduced pressure drop across columns which are quite stable and do not deteriorate. It is quite easy to understand that NPC is not suited for the separation of ionic and polar compounds. Very hydrophilic or ionic compounds are usually too strongly retained on polar sorbents with very less or no solubility in the organic MPs commonly used in NPC.

The chromatography that uses conditions in reverse to normal chromatography is known as reverse phase liquid chromatography (RPLC) [43]. RPLC uses a column containing highly non-polar modified silica produced by its reaction with a halogen-substituted organosilane. Thus, the modified silica may contain two, eight, or eighteen carbons, bonded at their ends through Si-O- Si groups to the surface of the support. Usually, octa C8 (8-carbon chain) and octadecyl C18 (18-carbon chain) are used. C8 column is less hydrophobic and less dense compared to the C18 column, leading to decreased retention times and reduced separation in C8 compared to the C18 column. Since the columns are highly non-polar in nature, the solutes (usually organic molecules) are retained due to dispersion within the medium.

Solvent delivery system is used for pumping solvents. The solvent is filtered to remove suspended particles to avoid damage to pumps and column plugging. The filtered solvent is then degassed by purging with helium to remove dissolved air to avoid disruption of detector signal. Vacuum filtration helps achieve both the pretreatments simultaneously. A HPLC pump should deliver a constant, pulse-free flow, and a reciprocating dual-piston pump is commonly used. Each piston moves in a small chamber with a very low volume of eluent, which is continuously passed through by check valves. When one piston slows, another one filled with MP pushes the MP, leading to pulse-free flow. Before shutting down the instrument, it is crucial to rinse the pump to remove salts, which will cause abrasion of pump components. Narrow-bore stainless steel tubing connectors are used as increased diameter or length of a tubing will affect the separation efficiency. Solvent gradient systems used for a mixture of solvent composition at low pressure. A six-port injector is used for high-pressure systems. These systems are equipped with a fixed volume loop of tubing that serves as the sample loop that is loaded with a standard blunt-tipped syringe. Sample loop sizes range from 1 to 100 µL for the analytical or preparatory scale. The advantage of an automatic sampler is that numerous samples can be automatically analysed by a computer-controlled system. An in-line filter is present immediately after the injection valve to remove any remaining particles in the MP to prevent clogging of the guard or analytical column. Guard columns are miniature versions of the analytical (separation) column containing the same material as SP. Guard columns typically cost one-fourth or less of the cost of an analytical column. The analytical column on which solutes are separated is very important and are available in different sizes, namely, preparatory column (20 – 50 mm in diameter X 50 – 250 mm in length), analytical column (4.5 mm in diameter X 12 – 25 mm in length), narrow-bore analytical columns (1 – 2 mm in diameter X 10 cm in length), capillary columns for MS detectors (from 0.075 to 0.1 mm in diameter). Larger columns require more volume of the MP to push the analytes through the system.

The choice of eluent is dependent on the nature of sample and the column to be used. In RPC, polar eluents like water and low-chain alcohols are used. The most polar solutes come out first from the column as they are least retained. Polarity index P' gives the strength of polarity of the MP. A higher value of P' indicates a more polar eluent. Generally, in separation, a mixture of solvents is used. The polarity index of the mixture M (P_M) containing solvents A and B can be calculated using individual polarity indices (P_A and P_B, respectively) and their respective volume fractions (V_A and V_B, respectively).

$$P_M = P_A * V_A + P_B * V_B \tag{7.4}$$

The effect of eluent polarity on the capacity factor k' of a solute is expressed mathematically by Eq. 7.5, wherein P1 and P2 are the polarity indices of two eluent mixtures. The eluent must be able to keep the sample components in solution. Viscosity of an eluent is also important because a less viscous solvent can be passed at a high flow rate without increasing the pressure to a great extent.

$$\frac{k'_2}{k'_1} = 10^{(P_2 - P_1)/2} \tag{7.5}$$

In RPC, the separation is carried out with a MP, e.g. a mixture of water and polar organic solvent like acetonitrile or methanol. This typically ensures the proper interaction of analytes with the non-polar, hydrophobic particle surface. A C18-bonded silica (ODS) is the most popular type of reversed-phase HPLC packing. The more polar analytes elute first, leaving the less polar analytes to be retained longer on the column.

The detection system is a very important part of instrumentation. In HPLC, a refractive index detector is considered to be a universal detector. But this detector needs excellent thermostat conditions and should be less selective and not suited for gradient elution. Ultraviolet absorption detectors are quite extensively used in the wavelength region of 210–900 nm. Below 210 nm, this is not suitable as most of the solvents would absorb in this spectral region. The sensitivity of the detector is found to be excellent. Fixed-wavelength detectors, PMT and photo diode array (PDA), are different types of detectors that can be used. A fixed-wavelength detector is not versatile as only compounds absorbing in that particular wavelength can be analysed. However, variable wavelength detectors with a continuum source are much more versatile. These detectors have a monochromator to select the desired wavelength. Detectors like PDA can rapidly scan over a range of wavelengths and so give both qualitative and quantitative information. In this detector, the diodes are arranged such that each diode intercepts different bands of wavelength. The flow detectors are designed in Z-shape to have a large path length within a small volume. Fluorescence detectors are quite sensitive but are limited to fluorescent compounds. Non-fluorescent compounds may be derivatized by adding a post-column reagent or may be determined by monitoring the reduction of the fluorescence of the eluent. This is known as "vacancy chromatography". Fluorescence detectors use a high intensity energy line (mercury lamp) or continuous source (deuterium or xenon arc) sources. A monochromator is used to select the wavelength for excitation and for emission, and a photomultiplier is used to amplify the weak emission signals. Fluorescence detectors are very sensitive but require very stringent conditions. The dissolved gases in solvents lead to quenching of signals, so degassing of solvents is important. Mass spectrometer is a very useful detector, but its interfacing with a HPLC system is difficult. Care should be taken that the solvent does not reach the vacuum system. Hence, the instrumentation is quite complex.

7.5.2 Ion chromatography

Ion chromatography is used for the separation of ionic species by using an ion exchanger resin as SP. The principles of ion exchange govern the separation [39]. Modern ion chromatography separations can follow any of the basic mechanisms of ion exchange, ion exclusion, and ion pair formation. IEC (commonly high-performance ion chromatography) is based on an ion-exchange process occurring between the MP and ion-exchange groups bonded to the support material (SP) (generally a polymer containing sulphonate or quaternary ammonium groups). Ion-exclusion chromatography (high-performance ion chromatography exclusion, HPICE) is steered by exclusion (Donnan and steric) and sorption using a sulphonated polystyrene/divinylbenzene-based cation ion exchanger as the SP. Ion-exclusion chromatography is useful for the separation of weak inorganic and organic acids (amino acids, aldehydes, and alcohols) from completely dissociated acids that are unretained and eluted without resolution along with void volume. Ion-pair chromatography (mobile phase ion chromatography, MPIC) uses neutral porous divinylbenzene resin or chemically bonded silica (octyl/octadecyl) of low polarity and high specific surface area as SPs. The selectivity is determined solely by the MP, which contains an ion-pair reagent which will react with the components of the sample to a different extent. It is suited for the separation of transition metal ion complexes.

Ion exchange is the process wherein ions in the liquid phase (electrolyte solution) stoichiometrically exchange with the labile ions of the insoluble solid phase (ion exchanger) containing fixed ionic sites. Cation exchange resins have fixed anionic sites and labile cations, while anionic exchange resins have fixed cationic sites and labile anions [44]. Amphoteric ion exchangers can exchange both cations and anions. It is very common that the labile cation in a cationic exchanger is a proton that is released in stoichiometric amounts when a cation in solution, say Na+, gets exchanged. Similarly, anionic exchangers have the labile OH anion, which can be released into a solution by different anions like chloride, iodide, etc. Ion exchangers possess a fixed surplus positive or negative charge (depending on the nature of resin AER/CER). This is charge neutralized by ions of opposite charge, known as counterions. The counterions move within the matrix and can be replaced by ions of same charge present in the solution. This is thought of as a sponge with the ions floating in pores [39]. When put in an ionic solution, the ions move in and out of the sponge in a stoichiometric ratio until equilibrium is reached. This simple model helps in understanding the electrostatic interactions but not the selectivity. Ion exchange is a stoichiometric reaction between the labile ions on the resin and similarly charged ions in the solution. Ion exchange resins classified as cation and anion exchange resins can be further classified as strong and weak resins depending the functional groups on the resin (the fixed ion). Apart from being classified into cationic and anionic, ion exchangers can be further classified into strong and weak. Resins having SO3H and COOH functional groups are classified as strong and weak cation exchange resins, respectively. The pH of operation for a strong resin is in the range of 0–14, while for a weak resin, the optimum pH is 7–14. The regenerant used for a cationic exchanger is acid. In case of an anionic exchanger resin, the presence of a tetrammonium group makes it a strong anion exchanger, which is useful in the high alkaline pH range of 10–14. The presence of a polyamine group makes the resin weak, and this resin can be used in the pH range of 1–7. Anionic exchanger resins are regenerated using a base.

Capacity is the parameter used to characterize the ion exchangers and is independent of experimental conditions. It is defined as the number of ionogenic groups present per specified amount of ion exchanger with units of meq/g of dry resin (H^+ and Cl^- for CER and AER, respectively). For a resin-packed column bed, technical volume capacity is expressed as eq/L. Apparent or effective capacity is the experimentally determined capacity that is dependent on experimental condition (pH and solution concentration) and is lower than maximum capacity. The selectivity coefficient, K, gives an idea of selectivity between ions and is dependent on various factors of the ions such as valency (higher valent ions preferred: tetra > tri > di > mono), ionic radius (the higher the radius, the less the hydration and higher the uptake), polarizability (the higher the polarizability, the better uptake), and complexing ability (the lower the tendency to form complexes, the higher the uptake).

SPs in ion chromatography contain organic polymers with good chemical stability. Different types of materials such as polymer-based, latex-agglomerated, and silica-based anion exchangers are used as SP. Styrene/divinylbenzene copolymers are favoured due to their stability in the pH range of 0–14. Latex-agglomerated anion exchangers are a special type of pellicular anion exchangers consisting of fully aminated porous polymer beads of high capacity, called latex particles (diameter 0.1 μm) agglomerated to the surface-sulfonated polystyrene/divinylbenzene substrate (with particle diameters 5–25 μm). The SP consists of three chemically distinct regions, namely, an inert and mechanically stable substrate, a thin coating of sulphonic acid groups over the substrate, and an outer layer of latex beads containing the anion exchange groups. Although the latex polymer has a very high exchange capacity (due to its complete amination), the small size decreases the ion exchange efficiency. The surface sulphonation thwarts the diffusion within the SP. These anion exchangers have better mechanical and chemical stability, faster ion exchange process, and better separation efficiency compared to silica-based anion exchangers and directly aminated resins. The high chromatographic efficiency of the separator column is attributed to the reduced swelling and shrinking of the small-sized latex beads owing to surface functionalization. Silica-based anion exchangers, contrary to organic polymers, have the advantages of higher chromatographic efficiency and greater mechanical stability even at elevated temperatures as there are no issues associated with the shrinking or swelling of particles with change in experimental conditions.

However, they have a restriction of use in a narrow pH range of 2–7. Macrocyclic SPs with electrically neutral macrocyclic compounds (crown ethers, cryptands, and calixarenes) can be used for separating anions (they form complexes with metal ions selectively, and anions will be part of the complex). The metal ion of the MP forms a complex with the macrocyclic compound, leading to a positive surface which can separate anions in the sample.

MPs generally are organic solvent buffers that have several specifications with respect to MP pH, buffer capacity, elution strength, complexing ability, counterion quality and concentration, and compatibility with the detection method. For the determination of anions and cations, multi-protic weak acids and bases, respectively, are added to the MP. The increase in pH of the MP increases its interaction with the SP, leading to displacement of ions from the SP. The pH maintained by the buffer should be as close as possible to that of the samples. Moreover, very high buffer concentrations can affect the retention of the ions on the SP and sometimes the complexing ability of the MP in the case of cation separation. The nature of the eluent used in anion chromatography is mainly dependent on the detection mode.

Chemical suppression is an important aspect of ion chromatography. Previous advances in suppressor functionality have focused on continuity, capacity, and high sensitivity, thus negating the need for chemical reagents for suppression. However, the use of a suppressor has become an imperative element in the anion analysis using conductivity detection. The suppressor is placed between the column and the detector to reduce the eluent background conductivity and also to enhance the conductivity of analyte ions. A high-capacity cation exchange membrane or resin in the acid form is used as a suppressor, which will remove the metal ions from both the eluent and the sample solution. This will not only reduce the background conductivity of the eluent but also enhance the signal of the analyte. The suppressor module consists of three cartridges filled with cation exchanger material, which will be sequentially placed prior to the analysis. The first cartridge is used for suppression, and at the same time, the second one is regenerated (with dilute sulphuric acid) and the third cartridge is rinsed with the eluate or water. Prior to each analysis, a freshly regenerated and rinsed cartridge is made available by rotation to 120°. Some instruments also have an electrolytically regenerated suppressor, which has features similar to that of a chemically regenerated suppressor. The only difference is that suppression is carried out by protons generated by water electrolysis. The use of hyphenated detection increases the back pressure placed on the suppressor, and the efficiency of newly developed packed columns also depends a great deal on the suppressor's efficiency. Thus, next-generation suppressors must have lower peak dispersion properties to support these columns. Previous generation electrolytic suppressors are very good at maintaining their regenerated form electrolytically, but if the suppressor loses its regenerated form (e.g. due to operation without current during installation), the performance suffers, and chemical regeneration is needed before again proceeding with electrolytic regeneration. Thus, future generation suppressors should be more current efficient and have sufficient static capacity to recover from this condition without the need for chemical regeneration. This is known as electrolytically regenerated suppressor, which has very high back pressure tolerance and is used in Reagent-Free™ Ion Chromatography (RFIC™) systems.

Detectors that are normally used ion chromatography, conductivity detectors, and UV-visible spectrophotometry are very commonly used. Fluorescence-based and electrochemical detectors are used for specific analysis. The other detection modes that can be combined with IC are atom absorption (AAS), inductively coupled plasma atomic emission spectrometer (ICP), and mass spectrometry (MS).

7.6 APPLICATIONS OF CHROMATOGRAPHY

Chromatography is considered as an omnipresent technique due to its versatility, simplicity, well-developed properties. The applications of chromatography are vast and often interdisciplinary. However, there is ongoing research on its applications in the analysis of various water bodies such as waste water and seawater. The great potential of chromatography in the analysis of water samples is due to the high-speed separations which are very simple. LC holds a special place in order to circumvent the limitations associated with other techniques (time-consuming, difficult to automate, low sensitivity, and poor selectivity due to spectral and chemical interferences). Ion chromatography is the most oft-used technique for the determination of ionic solutes. Ion chromatography was developed by Small et al. for the determination of alkali, alkaline earth, and ammonia cations using surface-sulphonated styrene–divinylbenzene copolymer resin.

The development of latex-coated DionexIonPac CS3 columns made it possible for the determination of ammonium ions [45] and different cations [46]. Analytical columns packed with iminodiacetic acid (IDA)-derivatized silica were used to separate alkali metal ions and ammonium ions in combination with non-suppressed conductivity detection [47]. The simultaneous separation of alkali and transition metals under isocratic conditions was achieved with an eluent comprising 10 mmol/l 18-crown-6, 1.5 mmol/l dipicolinic acid, and 1.9 mmol/l nitric acid. The chromatographic system enabled the quantitation of alkali metal ions with detection limits in the low parts per billion range and excellent linearity. Ion chromatography has been successfully used for the simultaneous determination of cations and anions. The simultaneous determination of cations and anions in one sample injection has been shown to depend on eluent selection [48]. Ion chromatography has scored over atomic spectrometric techniques due to its relatively low cost, ease of automation, online capability, and relatively wide dynamic range. Lanthanides have been determined by either cation or anion exchange chromatography using spectrometric detectors. Low- and high-molecular-weight amines were determined by either cation exchange or ion-pair chromatography, using conductivity detector in the suppressed or non-suppressed mode. The International Standard Organization (ISO) published Method 14911 for the simultaneous determination of dissolved alkali and alkaline earth cations, ammonia, and manganese in water and waste water is based on suppressed ion chromatography. Chromatography has become a well-established method for regulatory purposes as adopted by ISO, US EPA, and the American Society for Testing and Materials (ASTM) for environmental samples. ISO 14911 (1998) is a method for water quality monitoring involving the determination of Li^+, Na^+, NH_4^+, K^+, Mn^{2+}, Ca^{2+}, Mg^{2+}, Sr^{2+}, and Ba^{2+} in drinking water, waste water, and groundwater. ISO 10304-3 (1997) involves the determination of dissolved anions like chromate, iodide, sulphite, thiocyanate, and thiosulphate by in waste water. ISO 16749 (2005) is the IC method for the determination of hexavalent chromium in airborne particulate matter using spectrophotometric determination using diphenylcarbazide. US EPA Method 200.10 (1997) is used for trace element determination in marine water, brines, seawater, and estuarial waters for the determination of Cd^{2+}, Co^{2+}, Cu^{2+}, Pb^{2+}, Ni^{2+}, VO^{2+}, VO_2^{2+}, UO_2^{2+} by online chelation pre-concentration and inductively coupled plasma mass spectrometry detection. US EPA Method 200.13 (1997) gives the protocol for the determination of trace elements Cd^{2+}, Co^{2+}, Cu^{2+}, Pb^{2+}, Ni^{2+} in marine water, brines, seawater, and estuarial water by offline chelation pre-concentration with graphite furnace atomic absorption detection. US EPA Method 218.6 (1994) gives the standardized method for the determination of dissolved hexavalent chromium in drinking water, groundwater, and industrial effluent waste waters by ion chromatography. Different ASTM methods using ion chromatography have been developed. D5257 – 93 Standard Test Method is used for the determination of dissolved hexavalent chromium in drinking water, groundwater, and waste waters. D6504-00 standard practice is used for online determination of total conductivity of cations like Li^+, Na^+, NH_4^+, K^+, Mg^{2+}, and Ca^{2+} in high purity water. D6832-02 Standard Test Method is used for the determination of hexavalent chromium in workplace air by ion chromatography coupled with spectrophotometric measurement using 1,5-Diphenylcarbazide as the chromophore. D6919-03 Standard Test Method is utilized for the determination of alkali and alkaline earth cations and ammonium $\left(Li^+, Na^+, NH_4^+, K^+, Mg^{2+}, Ca^{2+} \right)$ in reagents, surface water, groundwater, and waste waters by ion chromatography. D6994-04 Standard Test Method is used for the determination of metal cyanide complexes $(Fe\,(CN)_6^{3-}$, $Fe(CN)_6^{4-}$, $Co(CN)_6^{4-}$, $Cu(CN)_6^{4-}$, $Ni(CN)_6^{4-}$, $Ag(CN)_6^{5-})$ in waste water, surface water, groundwater, and drinking waters using anion exchange chromatography coupled with UV detection. UOP 959-98 is a method for the determination of ammonium ions in drinking and waste waters using ion chromatography. WK653 is the method for the determination of dissolved alkali, alkaline earth cations, and ammonium in reagents, surface water, groundwater, and waste waters by ion chromatography. Dionex published methods for determination in various types of samples. AN4 is the method for the analysis of engine coolants, while AN25 is used for the determination of inorganic ions and organic acids in non-alcoholic carbonated beverages. AN69 and AN72 are the established protocols for Al in complex matrices and the determination of trace metals $\left(Cd^{2+}, Co^{2+}, Cu^{2+}, Fe^{2+}, Fe^{3+}, Pb^{2+}, Ni^{2+}, Mn^+, Zn^{2+}, VO^{2+}, VO_2^{2+} \right)$ in water-miscible organic solvents (water, alcohols, and acetonitrile) by ion chromatography/inductively coupled argon plasma spectroscopy (IC/ICAP), respectively. AN73 is the method for the determination of trace transition metals $\left(Cd^{2+}, Co^{2+}, Cu^{2+}, Pb^{2+}, Ni^{2+}, Mn^+, Zn^{2+}, Al^{3+} \right)$ in reagent-grade acids, bases, and their salts with IC/ICAP. AN77 gives the protocol for the determination of transition metals using chelation ion chromatography after the elimination of interferences of iron and aluminium. AN79 gives the protocol for the estimation of U and Th in complex matrix water samples, while AN80 is the method for the determination of dissolved hexavalent chromium in drinking water, groundwater, and industrial waste waters. Trace-level determination of cations in power plant samples and concentrated acids is given by protocols AN86 and AN94, respectively. The determination of alkali, alkaline earth, and ammonium ions in pharmaceuticals is given by AN106. AN108 is the protocol for the determination of transition metals in serum and whole blood by IC. The determination of Ca and Mg in brines is described in protocol AN120. The determination of transition metal ions in ppt levels in high purity water is given by AN131. Trace-level lithium determination in industrial process waters is illustrated in AU137. AN141 gives the details of the estimation of inorganic cations and ammonium in river, lake, ground, and sea waters using the IonPac CS16 column. AN144 is the methodology for the determination of hexavalent chromium in drinking water, while AN152 gives the details for the determination of Na (ppt) in the presence of high concentrations of ethanolamine in power plant waters. A reagent-free method is reported (AN155) for the determination of cations and amines in hydrogen peroxide. AN158 is the method for the determination of trace sodium and transition metals in power industry waste and reagent waters. TN24 and TN26 give methods for the determination of $Cr\left(Cr^{3+}, CrO_4^{2-}\right)$ by IC in wastewater and solid waste extracts. TN27 is the protocol for the determination of lanthanides $\left(La^{3+}, Nd^{3+}, Pr^{3+}, Ce^{3+}, Gd^{3+}, Tb^{3+}, Dy^{3+}, Tm^{3+}, Yb^{3+}\right)$ in digested rock samples by chelation ion chromatography. The most common application of ion chromatography is in the determination of alkali, alkaline earth metals, and ammonia. With the publication of ISO14911 Method, there was a surge in the utility of ion chromatography for cation analysis. All these methods used a wide range of separation columns, both commercially available (DionexIonPac CS12 and DionexIonPac CS12) and laboratory-made (IDA-bonded silica column). The eluents were also varied to a great range including EDTA, sulphosalicylic acid, and sulphuric acid, to name a few. Determination and, more specifically, the speciation studies of heavy, transition, and rare earth metal ions have become crucial due to the absence of simple and sensitive spectrometric methods. Therefore, hyphenated techniques based on chromatographic separation (IC) combined with highly selective and accurate detection modes (ICP-MS and ICP-AES) have gained more popularity [49, 50]. But the high cost is the main hurdle in the extensive use of hyphenated techniques. The determination of heavy metal ions requires complexation prior to separation in order to reduce hydrolysis of these metal ions and also to induce selectivity. The main care that needs to be taken is that the complexes do not precipitate within the column. Different ligands like oxalic acid, tartaric acid, citric acid, pyridine-2,6-dicarboxylicacid (PDCA), α-hydroxyisobutyric acid (HIBA), ethylenediaminetetraacetic acid (EDTA), etc., are used. Ion chromatography of metal–EDTA complexes can be carried out using anion-exchange columns enabling the separation of anions and cations $\left(Fe^{2+}, Pb^{2+}, Cu^{2+}, Zn^{2+}, Ni^{2+}, Co^{2+}, Cd^{2+}, Fe^{3+}\right)$ as anionic complexes in the same run [46–48]. It has also been observed that the interferences may be overcome by using chemically bonded phases [49, 50]. Different approaches including hyphenated techniques [51–54] have been used for the separation of metal ions. A large number of methods have been reported wherein different combinations of SP, eluents, and detectors have been used for the analyses of transition metal ions in a variety of samples [55–80]. The analysis of natural waters by ion chromatography alone covered about 99.96% of total cation and anion composition, and an excellent correspondence of the anion–cation balance was achieved. The use of various columns like AG4A, AS4A 1, AG5, AS5 1, AG12, CS12 1, Dionex, and IonPac in conjunction with eluents like sodium carbonate, bicarbonate, benzoate, citrate, methanesulphonic acid, oxalic acid, etc., have been utilized for analytical purposes using conductivity and spectrophotometric detectors.

In any method or instrumental development, the classical editorial of Herbert Laitinen on "The Seven Ages of an Analytical Method" [81], developed in analogy to Shakespeare's Seven stages of Man, should be kept in mind. According to this, the first stage is the initiation or conception phase [81], and the work carried out by H. Small was indeed the first stage in ion chromatography. When the work was awarded the Pittsburgh Applied Analytical Chemistry Award, the direction for the second stage was laid out to carry out a large number of experiments in different laboratories. The results were presented in the conference on Ion Chromatographic Analysis of Environmental Pollutants in 1978 [82]. This conference was dedicated "To the individual who does the best (s)he can with what (s)he's got" – best indicates the state of the field at this period [82]. The third stage is instrumentation development. In this stage, the method is brought into the hands of a non-specialist, and Model 10 Ion Chromatograph was demonstrated in the 1975 American Chemical Society (ACS) meeting in Chicago, IL, for which Dionex Corporation and Dow were awarded the 1977 Vaaler Award and 1977 IR100 Award [82]. The fourth stage involves detailed studies and improved instrumentation and ion chromatography matured as an analytical technique and various changes in both instrumentation and development of methods for specific samples were carried out [83]. It was in this stage that ASTM method for anions in water was first approved in 1984. IN 1898, an organization known as American Society for Testing and Materials (ASTM) was founded to improve product quality and personnel safety. ASTM test procedures are developed by experts for every industry. Thus the development of ASTM methods for anions was a mark of authentication of ion chromatographic analysis. Different ASTM methods were developed for determination of anions in water by suppressed ion chromatography (ASTM D 4327); for chloride, nitrate, and sulphate in atmospheric wet deposition by chemically suppressed ion chromatography (ASTM D 5085); and for dissolved hexavalent chromium in water by ion chromatography (ASTM D 5257). The determination of dissolved alkali, alkaline earth, and ammonium cations in drinking, ground, and municipal water samples has been reported (ASTM D 6919-032).

In the fifth and sixth stages, the applications become extensive into a wide range of fields with suitable modifications to the procedures. The seventh stage is the period of senescence when there are competing techniques that try to overshadow this method. Ion chromatography has undergone a tremendous development and offers an enormous range of possibilities for the selection of SPs and MPs, fabrication of novel separation modes, and hyphenation with different detection techniques. Ion chromatography can be carried out in 2D-IC and hyphenation techniques like ion chromatography-mass spectrometry (IC-MS) and in capillary mode resulting in capillary ion chromatography (CIC). High temperature IC is the new development based on the fact that at temperatures above 140°C, the dielectric constant of water is similar to that of hydrophobic organic solvents, and also the viscosity of a MP is decreased leading to reduced back pressure of the column. This leads to improvement of mass transfer within the column, enabling rapid separations with no peak tailings for complex separations. For anion exchange separations, the effect of temperature on selectivity can be quite complex coupled with the possible degradation of the column materials at above 60°C. Conversely, temperature offers significant opportunities for cationic separations as these columns are stable. Ion chromatography has been used for the analysis of nuclear materials. A recent review discussing the methodologies developed for the characterization of nuclear materials and related trace impurities using ion chromatography has been reported [84]. The studies used for monitoring of fuels, coolants, and control rods of a nuclear reactor have been discussed. Due to scientific and technical progression, modern instrumentation with elegant instrumentation, and efficient columns, IC finds extensive applications in various fields.

7.7 CONCLUSIONS

LC appears to be a versatile analytical technique. It separates the species present in a solution based on their affinities for the two phases. Ion chromatography is used for the separation of ions based on their affinities for an ion exchanger (SP). It has been observed that many detectors like AAS, multiple collector ICP-MS, thermal ionization mass spectrometry (TIMS), electrical conductivity, and UV detectors are used.

IEC is a popular purification method of proteins, peptides, nucleic acids, and other charged biomolecules, preferred for its high resolving power, high protein binding capacity, versatility with different types of ion exchangers, versatility with composition of buffers and pH, straightforward separation principle (primarily according to differences in charges), and ease of performance. IEC is a technique often used in protein purification, water analysis, and quality control, and it can be used for large proteins, small nucleotides, and amino acids. The principle of IEC is that charged molecules bind electrostatically to oppositely charged groups that have been bound covalently on the matrix. Wastewater samples can also be analysed using the ion chromatography as it has many advantages like good accuracy and precision, high selectivity, high speed, and separation efficiency. It is also used to develop the method, and the consumables are not expensive. But there is always a scope to improvise and develop new methods to improve the analysis parameters.

REFERENCES

1. J.G. Grasselli, *Mikrochim. Acta* 104 (1991) 545–549.
2. K. Cammann, *Fres. J. Anal. Chem.* 343 (1992) 812–813.
3. C. Burgess, *Valid Analytical Methods and Procedures*, The Royal Society of Chemistry, 2000, ISSN 978-0-85404-482-5, http://dx.doi.org/10.1039/9781847552280.
4. K. Danzer, *Fres. J. Anal. Chem.* 369 (2001) 397–402.
5. G. den Boef and A. Hulanicki, *Pure Appl. Chem.* 55 (1983) 553.
6. J. Vessman, R.I. Stefan, J.F. Van Staden, K. Danzer, W. Lindner, D.T. Burns, A. Fajgelj, and H. Müller, *Pure App. Chem.* 73 (2001) 1381–1386.
7. A. Fallon, R.F.G. Booth, and L.D. Bell (Eds.), Laboratory Techniques in Biochemistry and Molecular Biology, In *Chapter 1: The Origins and Development of Liquid Chromatography*, Vol. 17, pp. 1–7, Elsevier, 1987.
8. D.T. Day, *Proc. Amer. Philos. Soc.* 36 (1897) 112–115.

9. M.S. Tswett, *Berichte der Deutschenbotanischen Gesellschaft 24, 385* [translated and excerpted in Mikulás Teich, A Documentary History of Biochemistry, 1770–1940], Fairleigh Dickinson University Press, 1906 (1992), http://web.lemoyne.edu/~giunta/tswett.html.

10. A.J.P. Martin and R.L.M. Synge, *Biochem. J.* 35 (1941) 1358–1368.

11. I.D. Wilson, Chromatography| Paper Chromatography, In *Encyclopedia of Separation Science*, Ed. I.D. Wilson, pp. 397–404, Academic Press, 2000, ISBN 9780122267703.

12. H.G. Cassidy, *Anal. Chem.* 24 (1952) 1415–1421.

13. C.W. Jeanes and R.-J. Dimler, *Anal. Chem.* 23 (1951) 415–420.

14. L.S. Ettre and H. Kalász, *LC-GC N. Am.* 19 (2001) 712–721.

15. J.J. Fritz, *J. Chrom. A* 1039 (2004) 3–12.

16. C. Ó'Fágáin, P. Cummins, and B. O'Connor, *Methods Mol Biol (Clifton, N.J.)* 681 (2011) 25–33.

17. B.A. Adams and E.L. Holmes, *J. Chem. Soc.* 54 (1935) 1–6.

18. C.A. Lucy, J. Chrom. A 1000 (2003) 711–724.

19. S. Moore, D.H. Spackman, and W.H. Stein, *Anal. Chem.* 30 (1958) 1185–1190.

20. G.H. Lathe and C.R.J. Ruthven, *J. Polym. Sci.* 2 (1964) 835–843.

21. B.G. Belenkii and L.Z. Vilenchik (Eds.), *Chapter 4: Methodological Problems of Gel-Permeation Chromatography (GPC)*, Journal of Chromatography Library, Vol. 25, pp. 149–269, Elsevier, 1983.

22. K. Unger, *J. Chrom. A* 1060 (2005) 1–7.

23. L.R. Snyder and J.W. Dolan, Chapter 1 – Milestones in the Development of Liquid Chromatography, In *Liquid Chromatography*, Ed. S. Fanali, P.R. Haddad, C.F. Poole, P. Schoenmakers, and D. Lloyd, pp. 1–17, Elsevier, 2013, ISBN 9780124158078.

24. R. Meyer, Chromatography | Principles, In *Encyclopedia of Analytical Science*, Ed. P. Worsfold, A. Townshend, and C. Poole, 2nd ed., pp. 98–105, Elsevier, 2005.

25. S. Ahuja (Ed.), *Chapter 6: Chromatographic Methods*, Vol. 4, pp. 81–100, Separation Science and Technology, Academic Press, 2003.

26. O. Coskun, *North. Clin. Istanb.* 3(2) (11 November 2016) 156–160, https://doi.org/10.14744/nci.2016.32757.

27. W.J. Weber Jr., Y.-P. Chin, and C.P. Rice, *Wat. Res.* 20 (1986) 1433–1442.

28. N.S. Lakka and C. Kuppan, Principles of Chromatography Method Development, In *Biochemical Analysis Tools – Methods for Bio-Molecules Studies* [Internet], Ed. O.M. Boldura, C. Baltă, and N. Sayed Awwad, IntechOpen, 2020.

29. C. Rimmer, C. Simmons, and J. Dorsey, *J. Chrom. A* 965 (2002) 219–232.

30. M. Wang, J. Mallette, and J. Parcher, *J. Chrom. A* 1213 (2009) 105–109.

31. S. Ahuja, *J. Liq. Chromatogr.* 10 (1987) 1841–1845.

32. R. Snyder, *Chapter 1: Theory of Chromatography*, Ed. E. Heftmann, Journal of Chromatography Library, Vol. 51, Part A, pp. A1–A68, Elsevier, 1992.

33. JIS K0124:2011 General Rules for High Performance Liquid Chromatography, https://www.intertekinform.com/en-us/standards/jis-k-0124-2011-629855_saig_jsa_jsa_1455000/.

34. JIS K0214:2013 Technical Terms for Analytical Chemistry (Chromatography Part), https://www.intertekinform.com/en-us/standards/jis-k-0214-2013-1622679/.

35. L. Lapidus and N.R. Amundson, *J. Phy. Chem.* 56 (1952) 984–988.

36. J.J. van Deemter, F.J. Zuiderweg, and A. Klinkenberg, *Chem. Eng. Sci.* 5 (1956) 271–289.

37. F. Gritti and G. Guiochon, *J. Chromat. A* 1302 (2013) 1–13.

38. E. Kučera, *J. Chromatogr. A* 19 (1965) 237–248.

39. J.C. Giddings, *Dynamics of Chromatography*, Marcel Dekker, 1965.

40. U. Tallarek, F.J. Vergeldt, and H. Van As, *J. Phys. Chem. B* 103 (1999) 7654–7664.

41. P. Jandera, Chapter 1 – Comparison of Various Modes and Phase Systems for Analytical HPLC, In *Handbook of Analytical Separations*, Ed. Klára L. Valkó, Vol. 8, pp. 1–91, Elsevier Science B.V., 2020.

42. M.C. García-Alvarez-Coque, J.J. Baeza-Baeza, and G. Ramis-Ramos, Reversed Phase Liquid Chromatography, In *Analytical Separation Science*, Ed. V. Pino, J.L. Anderson, A. Berthod, and A.M. Stalcup, Wiley-VCH Verlag GmbH & Co. KGaA, 2015.

43. Jayshree Ramkumar and T. Mukherjee, Principles of Ion Exchange Equilibria, In *Ion Exchange Technology I*, Ed. I.M. Luqman, Springer, 2012.

44. R. Michalski and I. Kurzyca, *Pol. J. Environ. Studies* 15 (2006) 5–18.

45. R. Michalski, *Crit. Rev. Anal. Chem.* 39 (2009) 230–250.

46. J. Weiss and D. Jensen, *Anal. Bioanal. Chem.* 375 (2003) 81–98.

47. Paul R. Haddad (Ed.), *Chapter 3 Ion-Exchange Stationary Phases for Ion Chromatography*, Journal of Chromatography Library, Vol. 46, pp. 29–77, Elsevier, 1990.

48. R.Q. Yang, et al., *Talanta* 55(6) (2001) 1091–1096.

49. A. Haidekker and C. Huber, *J. Chromatogr. A* 921 (2001) 217–226.

50. R. Dumanli, A. Attar, V. Erci, and I. Isildak, *J. Chrom. Sci.* 54 (2016) 598–603.

51. J.G. Tarter, *J. Chromatogr. Sci.* 27 (1989) 462–467.

52. L.A. Ellis and D.J. Roberts, *J. Chromatogr. A* 774 (1997) 3–19.

53. M. Montes-Bayon, K. DeNicola, and J.A. Caruso, *J. Chromatogr. A* 1000 (2003) 457–476.

54. T.Z. Tanaka, *Anal. Chem.* 320 (1985) 125–127.

55. P. Hajos and G. Revesz, *J. Chromatogr.* 640 (1993) 15–25.

56. P. Hajós, G. Révész, O. Horváth, J. Peear, and C. Sarzanini, *J. Chromatogr. Sci.* 34 (1996) 291–299.

57. S. Chandramouleeswaran, B. Vijayalakshmi, S. Kartihkeyan, et al., *Mikrochim. Acta* 128 (1998) 75–77.

58. T.P. Rao, J. Metilda, and M. Gladis, *Crit. Rev. Anal. Chem.* 35 (2005) 247–288.

59. R.K. Sharma, S. Mittal, and M. Koel, *Crit. Rev. Anal. Chem.* 33 (2003) 183–197.

60. T. Guierin, A. Astruc, and M. Astruc, *Talanta* 50 (1999) 1–24.

61. H. Collins, S.H. Pezzin, J.F.L. Rivera, P.S. Bonato, C.C. Windmöller, C. Archundia, and K.E. Colins, *J. Chromatogr. A* 789 (1997) 469–478.

62. Y.J. Chen, L. Jing, X.L. Li, and Y. Zhu, *J. Chromatogr. A* 1118 (2006) 3–11.

63. W.F. Zeng, Y.X. Chen, H.R. Cui, F.Y. Wu, Y. Zhu, and J.S. Fritz, *J. Chromatogr. A* 1118 (2006) 68–72.

64. N. Cardellicchio, P. Ragone, S. Cavalli, and J. Riviello, *J. Chromatogr. A* 770 (1997) 185–193.

65. E. Sugrue, P.N. Nesterenko, and B. Paull, *Anal. Chim. Acta* 553 (2005) 27–35.

66. E. Santoyo, S. Santoyo-Gutiérrez, and S.P. Verma, *J. Chromatogr. A* 884 (2000) 229–241.

67. D.T. Gjerde, *J. Chromatogr. A* 439 (1988) 49–61.

68. E.H. Borai and A.A. Soliman, *J. Chromatogr. A* 920 (2001) 261–269.

69. P.L. Buldini, S. Cavalli, A. Mevoli, and J.L. Sharma, *Food Chem.* 73 (2001) 487–495.

70. P. Bruno, P. Caselli, G. de Gennaro, P. Ielpo, and A. Traini, *J. Chromatogr. A* 888 (2000) 145–150.

71. H.T. Lu, S.F. Mou, and J.M. Riviello, *J. Chrom. A* 857 (1999) 343–349.

72. A. Rahmalan, M.Z. Abdullah, M.M. Sanagi, and M. Rashid, *J. Chrom. A* 739 (1996) 233–239.

73. T. Williams, P. Jones, and L. Ebdon, *J. Chromatogr. A* 482 (1989) 361–366.

74. R. Michalski, *J. Liq. Chrom. Rel. Technol.* 28 (2005) 2849–2862.

75. M. Pantsar-Kallio, K. Pentti, and M.G.J. Manninen, *Chromatogr. A* 779 (1997) 139–146.

76. N. Ulrich, *Anal. Chim. Acta* 359 (1998) 245–253.

77. A.F. Roig-Navarro, Y. Martinez-Bravo, F.J. Lopez, and F.J. Hernandez, *Chromatogr. A* 912 (2001) 319–327.

78. R.T. Gettar, R.N. Garavagila, E.A. Gautier, and D.A.J. Batistoni, *Chromatogr. A* 884 (2000) 211–221.

79. A.W. Al Shawi and R. Dahl, *Anal. Chim. Acta* 333 (1996) 23–30.

80. H.A. Laitinen, *Anal. Chem.* 45 (1973) 2305.
81. E. Sawicki, J.D. Mulik, and E. Wittgenstein (Eds.), *Ion Chromatographic Analysis of Environmental Pollutants*, Ann Arbor Science, 1978.
82. E. Sawicki, J.D. Mulik, and E. Wittgenstein (Eds.), *Ion Chromatographic Analysis of Environmental Pollutants*, Ann Arbor Science Publishers, Inc., pp. vi + 210, 1978, *J. Chr. Sci.* 16 (1978) 18A, https://doi.org/10.1093/chromsci/16.5.18A-c.
83. R. Michalski, *Crit. Rev. Anal. Chem.* 36 (2006) 107–127.
84. H. Parab, Jayshree Ramkumar, A. Dudwadkar, and S.D. Kumar, *Rev. Anal. Chem.* 40 (2021) 204–219.

8 Broad conclusions and future perspectives

Analytical chemistry is known to be the basis of all studies. In this field, separation and sensing are two most important aspects. Separation science plays a crucial role in our daily life. It not only helps us in carrying out our daily activities but also helps in maintaining the quality of our life. The incessant growth of industries has made it crucial for the development of separation processes, not only for purification but also for monitoring of various important species. The role of separation also becomes very important in the overgrowing need for remediation of environmental media. It is very true that the field of separation science is ever growing due to the new challenges that arise and also to fill in the gaps of the present challenges. Thus, the field of separation science could be used to achieve improved living for mankind and also have efficient production, thus reducing costs.

The need for providing sustainable good quality water makes it necessary to adopt remedial techniques and also have a set of standards not only for the specifications of potable water but also for the reusability of water. For the past many years, reuse of water has become an important requirement, and therefore, separate guidelines are needed. EPA has developed the 2012 Guidelines for Water Reuse to incorporate this information through a Cooperative Research and Development Agreement (CRADA) with CDM Smith and an interagency agreement with US Agency for International Development (USAID). This shows the ever-growing research on the advances in methodologies adopted for water reuse. With the presence of excellent methodologies, it is possible to treat waste water and produce water deemed fit for use. Thus, an integrated approach of water management is adopted, wherein non-conventional water sources are also considered and the findings of various regulatory bodies [National Research Council's (NRC) Water Science and Technology Board report, Water Reuse: Potential for Expanding the Nation's Water Supply Through Reuse of Municipal Waste water (NRC, 2012)] are integrated. Similarly, WHO has also updated its guidelines. The guidelines can prove beneficial to those involved in various aspects of water management, reclamation, or reuse. In such a study of interdisciplinary nature, the knowledge of various technical terms becomes very crucial. De facto reuse is when the reuse is being practised but not officially recognized. Direct potable reuse (DPR) is the introduction of remediated water into drinking water treatment plants. Indirect potable reuse (IPR) refers to the augmentation of a drinking water source (surface or groundwater) with reclaimed water followed by an environmental buffer that precedes drinking water treatment. Non-potable reuse refers to all other applications except that used for potable reuse (drinking purposes). There are different types of applications of reuse of reclaimed water. In the urban applications, the use of reclaimed water for non-potable applications in municipal settings is known as unrestricted or restricted depending upon whether public access,

DOI: 10.1201/9781003442516-8

respectively. Public access can be controlled or restricted by physical or institutional barriers (fencing or advisory signage). The water can be used for agricultural reuse for irrigation of food crops and also other types of crops. The main motivation for reuse is to augment the water resources to cope up with the ever-increasing demands for water. The various benefits of reuse of water involve improved agriculture produce, decreased energy consumption of treatment processes, reduced nutrient loads, etc. The main focus of all this integration is to address water supply scarcity, efficient resource use, and environmental and public health protection. However, the key to all these developments is sustainability. As per the Brundtland Commission report, "Sustainable development is development that meets the needs of the present without compromising the ability of future generations to meet their own needs"(WCED, 1987). Therefore, sustainable water management involves management of the present resources to have no strain on future generations. It is now well understood by environmentalists that both water and energy being interconnected make it essential to consider each other. Therefore, the practice of water reuse can not only offset water demands but also provide additional water for energy production. EPA has developed principles for an energy–water future which emphasize on the use of waste water or reusable water as a resource (EPA, 2012). Since water reuse is a great opportunity, EPA has developed a handbook entitled *Leveraging the Water–Energy Connection – An Integrated Resource Management Handbook for Community Planners and Decision-Makers*, as a manual to be followed. The manual will address water conservation and efficiency; alternative water sources, building codes for improved water and energy use efficiency; and renewable energy sources from/for both water and wastewater systems. In India, it is only in recent times that water and wastewater sectors, especially wastewater reuse, have started attracting the attention of many. However, many challenges like poor infrastructure and low social acceptance need to be overcome to make the concept of water reuse a reality. The Central Public Health & Environmental Engineering Organisation (CPHEEO) estimates indicate that about 70–80% of total water supplied for domestic use gets generated as waste water. As per CPCB estimates, the total wastewater generation from class I cities (498) and class II (410) towns in the country is around 35,558 and 2,696 MLD, respectively, while the installed sewage treatment capacity is 11,553 and 233 MLD, respectively, thereby leading to a gap of 26,468 MLD in sewage treatment capacity. It is known that states like Maharashtra, Uttar Pradesh, West Bengal, and Gujarat and Delhi contribute to the major share (63%) of waste water (CPCB, 2007a), and reports have shown that the industrial water use productivity of India is lowest. It is projected that water scarcity would increase tremendously by 2050, and there would be a concurrent depletion of freshwater resources and also increased wastewater generation. Thus, treated waste water can be used for irrigation of crops, maintaining parks and golf courses, etc. Research on efficient, cost-effective, and sustainable treatment protocols needs to be encouraged, and also the concept of phytoremediation should be more seriously adopted. It is also important to have in-depth knowledge of indigenous or traditional methods, and the methods should be easily available for further research.

It is to be understood that wastewater reuse has still not gained lot of attention due to the associated problems and challenges. The great challenge is the development of economically viable and easy-to-use techniques without causing huge changes in the environmental system. The use of constructed wetlands as a means of phytoremediation is catching up due to inherent advantages like energy efficiency, ease of operation, no secondary waste, and no need of highly qualified technical staff to carry out the treatment. Thus, it is seen that wastewater treatment can not only provide a healthy environment but can also increase the scope for reuse to overcome water scarcity issues.

Water resource management is a complex process that not only looks for treatment of waste water but also plays a major role in its sustenance. It is indeed a challenge due to the increase in the demands and also the deterioration of water quality. So there are different types of approaches involved. Unidisciplinarity is when a researcher is from a single discipline who tends to work alone or collaboratively to address a common question, problem, topic, or theme. Multidisciplinarity involves the focus of different disciplines on a particular problem, resulting in an array of information, knowledge, and methods, but disciplines remain separate. The existing structure of knowledge is not questioned. Interdisciplinarity leads to the integration of information, data, methods, tools, concepts, and/or theories from two or more disciplines focused on a complex question or problem. The key defining concept of interdisciplinarity is integration of diverse inputs. The team would draw on multiple, integrated data sources generated through the lens of interdisciplinary theory and use of interdisciplinary methods. Transdisciplinarity leads to the transcending of disciplinary approaches using more comprehensive frameworks, including synthetic paradigms. The construct goes beyond interdisciplinary combinations of existing approaches and fosters new worldviews or domains. Thus, water management requires extensive know-how of hydrological aspects and conservation and treatment procedures, and should most importantly be able to implement and integrate strategies to make it a community and then a global effort. Researchers have suggested an interdisciplinary approach to addressing challenges to water-related issues. Since the issues are complex, an interdisciplinary approach was often needed to effectively confront these challenges. Interdisciplinary research involves researchers with very different backgrounds to understand and communicate with each other. But such collaborations may lead to the need to cross one's own field of expertise, which proves to be a stumbling block in the success of such collaborations. The apprehensions and challenges can be easily understood from the words of Hannes Alfven, "Scientists tend to resist interdisciplinary inquiries into their own territory. In many instances, such parochialism is founded on the fear that intrusion from other disciplines would compete unfairly for limited financial resources and thus diminish their own opportunity for research". But in the long run, this will be useful as pointed out by Robert J. Shiller, who said "In the longer run and for wide-reaching issues, more creative solutions tend to come from imaginative interdisciplinary collaboration".

When it comes to water, a very famous quote by Poet Samuel Taylor Coleridge (1772–1834) in *The Rime of the Ancient Mariner* (1798) goes: "Water, water, everywhere, And all the boards did shrink; Water, water, everywhere, Nor any drop to drink". It can be understood that despite having the resources, you cannot benefit from it. In the case of water systems, this is directly related and conveys the fact that the level of pollution has made it difficult to use the resources, and in common usage, the above line is usually misquoted as "Water, water everywhere, but not a drop to drink". In this era, this is an intimidating reservation that scientists and technologists and various other experts are trying to address. This intensive collaboration aims not only to remove the toxic species from water bodies but also to reduce the discharge of these effluents. As there are a large number of techniques that are developed and still developing, it is only indicating how important the issues of water pollution and subsequent remediation are.

In the end, it is only justifiable to quote Juan Evo Morales Ayma, a Bolivian politician, who said "Sooner or later, we will have to recognise that the Earth has rights, too, to live without pollution. What mankind must know is that human beings cannot live without Mother Earth, but the planet can live without humans". Therefore, the combined efforts of all the people is the only key to it, as Margaret Mead, an American cultural anthropologist, pointed out: "Never doubt that a small group of thoughtful, committed citizens can change the world; indeed, it's the only thing that ever has".

Index

Note: Page numbers in *italics* indicate a figure on the corresponding page.